THE ANIMAL BODY BOOK

JESS FRENCH

Illustrated by **Jonathan Woodward**

Author Dr Jess French
Illustrator Jonathan Woodward

Senior editor Olivia Stanford
Project art editor Polly Appleton
Additional design Sonny Flynn, Hannah Moore
Managing editor Gemma Farr
Managing art editor Diane Peyton Jones
Senior production editor Nikoleta Parasaki
Senior production controller Leanne Burke
Picture research Nunhoih Guite, Vagisha Pushp
Jacket coordinator Elin Woosnam
Jacket designer Elle Ward
Art director Mabel Chan
Managing director Sarah Larter

Consultant Dr Catrin Rutland

First published in Great Britain in 2024 by
Dorling Kindersley Limited
DK, One Embassy Gardens, 8 Viaduct Gardens,
London, SW11 7BW

The authorised representative in the EEA is
Dorling Kindersley Verlag GmbH. Arnulfstr. 124,
80636 Munich, Germany

Text copyright © Jess French 2024
Copyright © 2024 Dorling Kindersley Limited
A Penguin Random House Company
10 9 8 7 6 5 4 3 2 1
001–336956–June/2024

All rights reserved.
No part of this publication may be reproduced,
stored in or introduced into a retrieval system,
or transmitted, in any form, or by any means
(electronic, mechanical, photocopying, recording,
or otherwise), without the prior written
permission of the copyright owner.

A CIP catalogue record for this book
is available from the British Library.
ISBN: 978-0-2416-3526-1

Printed and bound in China

www.dk.com

CONTENTS

INTRODUCTION 4
STUDYING ANATOMY 6
BODY SHAPES 8
VERTEBRATES 10
ORGAN SYSTEMS 12
COMPARATIVE ANATOMY 14

SKELETONS

WHAT IS A SKELETON? 18
CARTILAGE 20
SKULL 22
FUSED BONES 24
PNEUMATIC BONES 26
HANDS 28
INSIDE A BAT 30

MUSCLES

WHAT IS MUSCLE? 34
PAIRED MUSCLES 36
TENDONS 38
CHEWING 40
GRASPING BODY PARTS 42
INSIDE A CHAMELEON 44

CIRCULATION

WHAT IS CIRCULATION? 48
HEART 50

SINGLE CIRCULATION.................52
BLOOD...................................54
TEMPERATURE........................56
INSIDE A GIRAFFE...................58

RESPIRATION

WHAT IS RESPIRATION?............62
GILLS....................................64
BIRD LUNGS..........................66
MAMMAL LUNGS...................68
SOUND.................................70
INSIDE A SNAKE....................72

DIGESTION

WHAT IS DIGESTION?..............76
MOUTH................................78
STOMACH.............................80
LIVER...................................82
INTESTINES..........................84
KIDNEYS...............................86
TEETH..................................88
INSIDE A PARROT..................90

SENSES

WHAT ARE SENSES?................94
EYES....................................96
NOSE...................................98
EARS..................................100
WHISKERS...........................102
EXTRA SENSES....................104
INSIDE AN ELEPHANT...........106

REPRODUCTION

WHAT IS REPRODUCTION?......110
EGGS WITHOUT SHELLS........112
EGGS WITH SHELLS..............114
FEMALE ANATOMY...............116
MALE ANATOMY..................118
ORNAMENTS.......................120
INSIDE A KANGAROO............122

INTEGUMENT

WHAT IS INTEGUMENT?.........126
BLUBBER............................128
SKIN..................................130
SCALES..............................132
FEATHERS..........................134
FUR...................................136
INSIDE A SHARK..................138

INVERTEBRATES

WHAT IS AN INVERTEBRATE?...142
COMPOUND EYES.................144
SPIRACLES.........................146
PUPA.................................148
TENTACLES........................150
ARMOUR............................152
INSIDE AN OCTOPUS............154

GLOSSARY..........................156
INDEX................................158
ACKNOWLEDGEMENTS.........160

INTRODUCTION

I have been fascinated by animal bodies for as long as I can remember, from collecting dry, dusty bones I found on walks in the countryside to peering intently into my pets' mouths and ears to learn more about the way they worked. Now, in my job as a vet, understanding animal bodies is a vital part of how I help them to get better.

Animals are all different, and just when I start to think I've learned all there is to know about animal anatomy, I learn a brand new fact that blows me off my feet! That's what I love about science – there's always more to discover…

Join me to delve beneath the fur, feathers, and scales of some of the world's most extraordinary animals as we learn about how they work, from the inside out!

Jess French

JESS FRENCH

ANIMAL ANATOMY

DISSECTIONS

One of the simplest ways to study animal anatomy is to look at it directly! When scientists cut open a deceased animal to see its insides, it is known as dissection. By dissecting an animal, such as a frog, the shape, type, and number of its organs can be studied.

Frog dissection

The skin of the frog is cut and moved to the side, so the organs can be seen easily.

A dissection allows scientists to see the size and location of the internal organs.

A very sharp knife called a scalpel is used for dissections.

STUDYING ANATOMY

Every animal is different inside, so how do we find out what their organs look like? There are lots of different ways. For many years, all scientists could do was to cut animals open once they were dead, to look directly at their insides or to study their **bones**. Today, there are many types of scans that allow us to see inside animals while they are still alive and which can show us how **organs** work too.

STUDYING SKELETONS

Once an animal has died, its soft **tissues** quickly break down, leaving behind just its **skeleton**. Scientists can study animal skeletons to see what their bones look like and what they are used for. However, skeletons can only tell us about one part of an animal's anatomy.

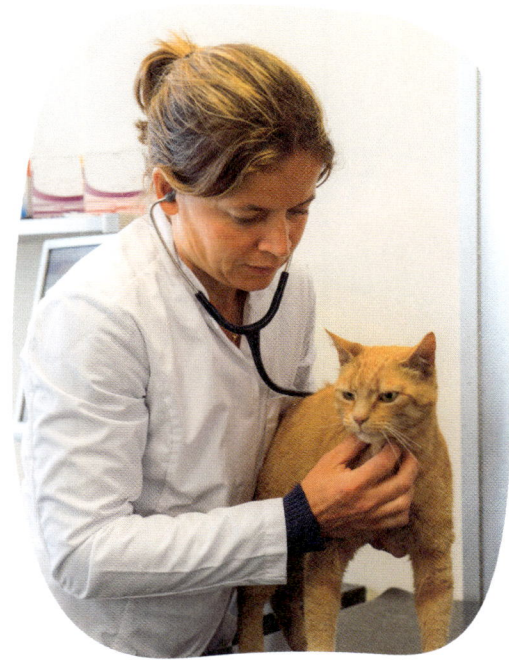

VETS

Vets must understand the anatomy of all sorts of animals to be able to treat them when they become ill. They spend many years learning about animal anatomy, so that when they come to perform surgery, they know how and where to operate.

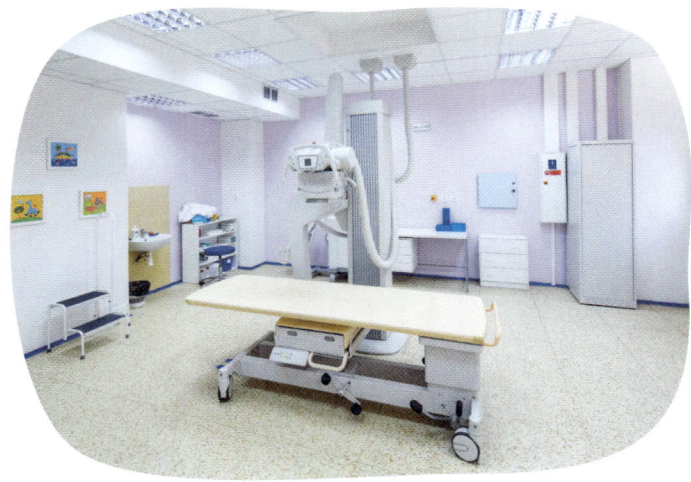

IMAGING

There are now many ways of seeing inside an animal without having to cut it open. X-rays allow us to see the bones and hard parts of an animal, and special scans called MRIs, CTs, and ultrasounds can create images that let us see soft tissues and organs too.

X-ray image of a swan's head and neck

ANIMAL ANATOMY

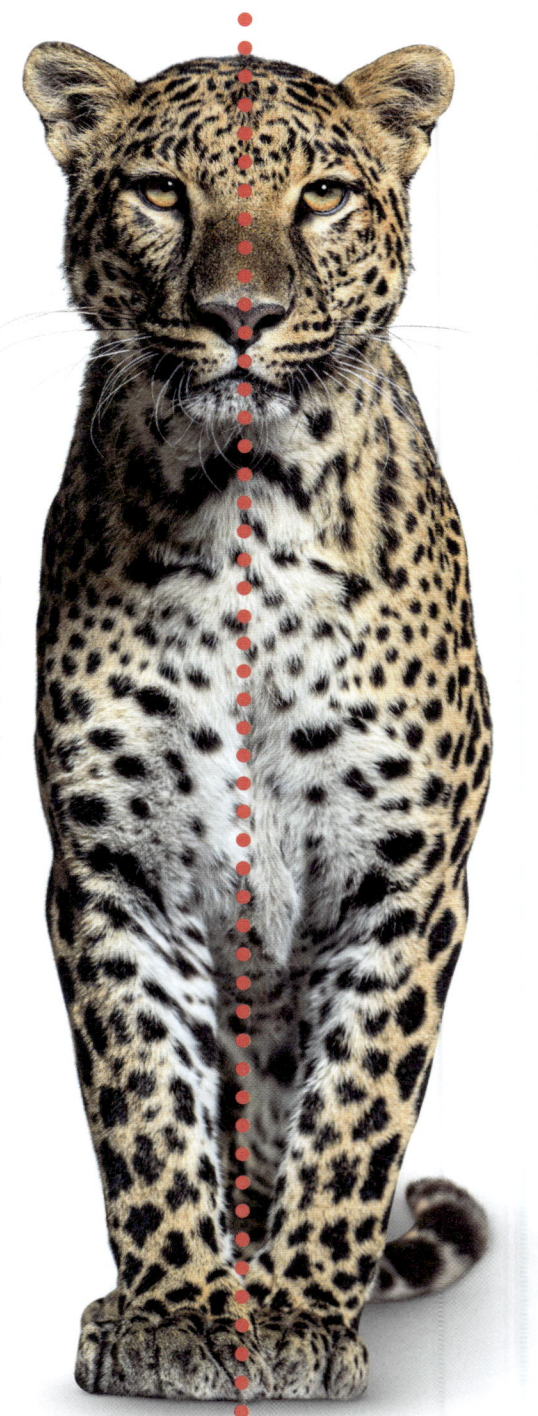

SYMMETRY

On the outside, most animals look symmetrical in some way, even if they aren't on the inside. This means they look the same when reflected or rotated along imaginary lines. Not all animals are symmetrical though, such as snails with spiral shells on one side of their body. We call these animals asymmetrical.

Radial symmetry
Many invertebrates have radial symmetry. This means if you imagined rotating the animal around its centre, it would make the same shape more than once. Starfish with five arms have five lines of symmetry.

Bilateral symmetry
Most animals have bilateral symmetry. This means that if you imagined a line down their centre, they would look like they were reflected on either side. Internally, however, many animals are asymmetrical, with **organs** such as the stomach to one side of the body.

BODY SHAPES

Animals come in a huge variety of shapes, from five-armed starfish to four-legged leopards, and everything in between. Often, members of the same animal group have similar-looking bodies, but the **environment** a **species** lives in can affect its shape too. Even when species look very different from one another, their shapes often follows certain rules.

ANCIENT ANATOMY

Fossils tell us about the bodies of ancient animals, and many looked completely different to animals alive today. From dinosaurs to woolly mammoths, many prehistoric animals were much bigger than those we are familiar with. One relative of the elephant, Platybelodon, had an unusual long lower jaw and tusks to strip bark off trees.

PAIRED SENSE ORGANS

Most animals with eyes, ears, or nostrils have at least two of each, which helps them tell which direction a sight, sound, or smell came from. The eyes of **predators** often both face forwards to help them judge distance better. However, the two eyes of **prey** animals are usually on the sides of their heads so they can see all around.

CHANGING ANATOMY

All animals change their anatomy as they grow. Young animals are sometimes a different colour or pattern to adults and have different proportions. Some animals change very dramatically as they grow, such as tadpoles transforming into frogs or animals that look very different depending on their sex, such as female and male orangutans.

ANIMAL ANATOMY

VERTEBRATES

Many large animals are **vertebrates**, which means they have a **skeleton** inside their body that includes a **backbone**. This backbone can be made of **bone** or **cartilage**, depending on the animal. There are five main groups of vertebrates: **fish**, **amphibians**, **reptiles**, **birds**, and **mammals**.

Reptile
Reptiles have dry, scaly skin and most of them live on land. Some lay eggs with soft or hard shells to reproduce, but others give birth to babies.

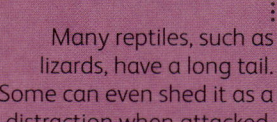

Many reptiles, such as lizards, have a long tail. Some can even shed it as a distraction when attacked.

Tiny, hard plates called scales grow from the skin of most fish. They can be brightly coloured.

Instead of arms or legs, fish have thin fins to help them swim.

Almost all fish breathe using **organs** called gills, found behind the head.

Fish
Most fish live underwater their whole lives. They have fins to help them swim and **gills** so they can breathe. Many have a long, pointed body covered in hard scales that give them a streamlined shape.

When young, amphibians often have gills to breathe underwater, but develop **lungs** to breathe air as adults.

Thin, moist skin allows amphibians to absorb **oxygen** directly into their body.

Many amphibians, including salamanders, have a long tail. Frogs and toads, however, only have tails when young.

Most amphibians have four legs.

Amphibian
Amphibians live for part of their lives on land and part in water. They have thin, moist skin, and most have four limbs to clamber around. They lay eggs to reproduce, which must be kept damp while they develop.

All reptiles have hard scales covering their body.

Unlike amphibians, reptiles have claws on the ends of their toes.

Most reptiles have four legs, but some, such as snakes, don't have any legs.

Every mammal has some hair. Many mammals are completely covered in it.

Most mammals have four limbs. Some, such as cows, walk on four legs.

All mammals feed their young on milk, which is produced in special **glands**.

Mammal

Mammals are recognizable by their hair or fur, which helps to keep them warm. Almost all reproduce by giving birth to babies, which are fed on milk when young.

A bird's feathers can help to keep it warm and waterproof.

Birds gather food using their beak, also called a bill.

Birds have scaly legs and some, such as ducks, have webbed feet to help them swim.

Bird

Two key characteristics of birds are feathers, which allow many species to fly, and a hard beak to collect food with. All birds lay hard-shelled eggs to reproduce, and many build a nest to keep their eggs safe.

INVERTEBRATES

Animals without a backbone are called **invertebrates**. There are many more types of invertebrate then vertebrate. Some have hard outer shells instead of a skeleton, such as shellfish, or tough external armour, like insects. Many have no hard support for their body at all, such as worms and octopuses.

ANIMAL ANATOMY

ORGAN SYSTEMS

There are lots of jobs to do inside an animal's body! For animals to grow, move, and thrive, their bodies must complete many different tasks – this is the work of special structures called **organs**. Groups of organs that work together to perform a specific function, such as making an animal move or digesting food, are called systems. All the systems work together to keep an animal alive.

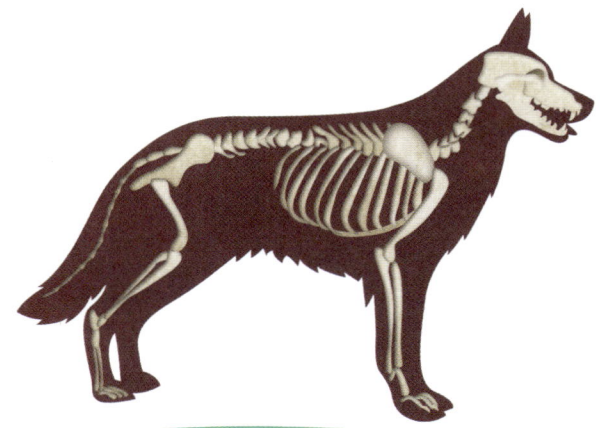

Skeletons
All **vertebrates** have a **skeleton**, which is made from hard **bones** or flexible **cartilage**. It provides a sturdy frame that gives an animal its shape and also provides an attachment point for an animal's **muscles**.

Muscles
Muscles are what allow an animal to move. When muscles shorten, or contract, they pull on the bones of the skeleton to reposition them. Some organs, such as the **heart**, are also made from muscles.

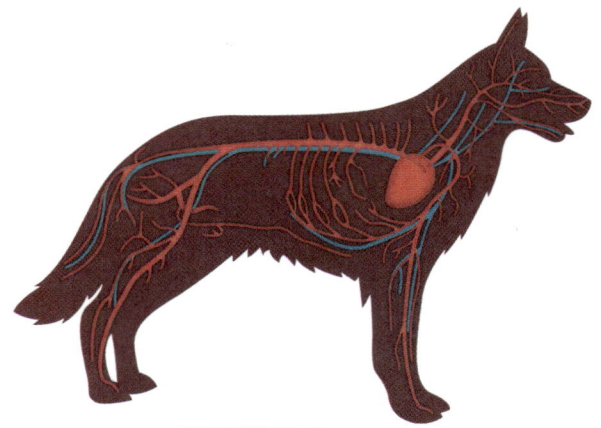

Circulation
Vertebrates need a system to deliver vital **nutrients** and **oxygen** to their organs. This is done by **blood**, which is pumped around a network of tubes by the heart. Together these are known as the circulatory system.

Respiration
Animals need oxygen to survive, and they can extract it from either water or air depending on whether they have **gills** or **lungs**. The oxygen is used in a **chemical reaction** called **respiration**, which generates energy for the body to use.

UNUSUAL ORGANS

Invertebrates have very different body systems to vertebrates. Many have completely different organs, but they must perform the same functions as those of vertebrates. The flower-like gills of sea slugs are one of many specialized organs found in different groups of invertebrates.

Digestion
To get the nutrients they need, vertebrates eat food. However, it must be broken down to be used. This is the job of the digestive system. The mouth and stomach liquefy food and the useful parts are absorbed in the **intestines**.

Senses
Animals must be able to sense the world around them and respond to it in order to survive. This is the work of the nervous system, which contains a network of wirelike nerves that carry messages between the body and the **brain**.

Reproduction
All species must reproduce in order for them to survive. Producing young is the task of the reproductive system. Usually, male and female vertebrates use special organs to produce different **cells** that, when joined, create a new baby animal.

Integument
All vertebrates have skin, a protective covering that keeps germs out and water in. Over the skin, many animals have extra coverings, such as hair, feathers, or scales. Together, the skin and these extra coverings are known as integument.

COMPARATIVE ANATOMY

One of the most interesting areas of animal anatomy is looking at the differences between different **species** to see how they are **adapted** for their **environments**. Comparing anatomy between animals – which is called comparative anatomy – can teach us a lot about how our own bodies work and even give us new ideas for science and technology.

COMPARING ANIMALS

Different animals have **evolved** in different ways, but animals from related groups often have many similarities. Birds and bats are both **vertebrates** and have the same types of **bones** in their wings, but if we look at them side by side, those bones are completely different shapes.

A streamlined body is essential to move through the water easily. Sharks and dolphins have evolved a smooth, cylindrical body to cut through the water.

HABITATS

One of the main factors that affects an animal's anatomy is its habitat. For example, animals that live in cold polar regions, such as the Arctic fox, often have thick fur for insulation and small ears to prevent heat loss. Even animals that aren't closely related can look similar when they live in the same habitat.

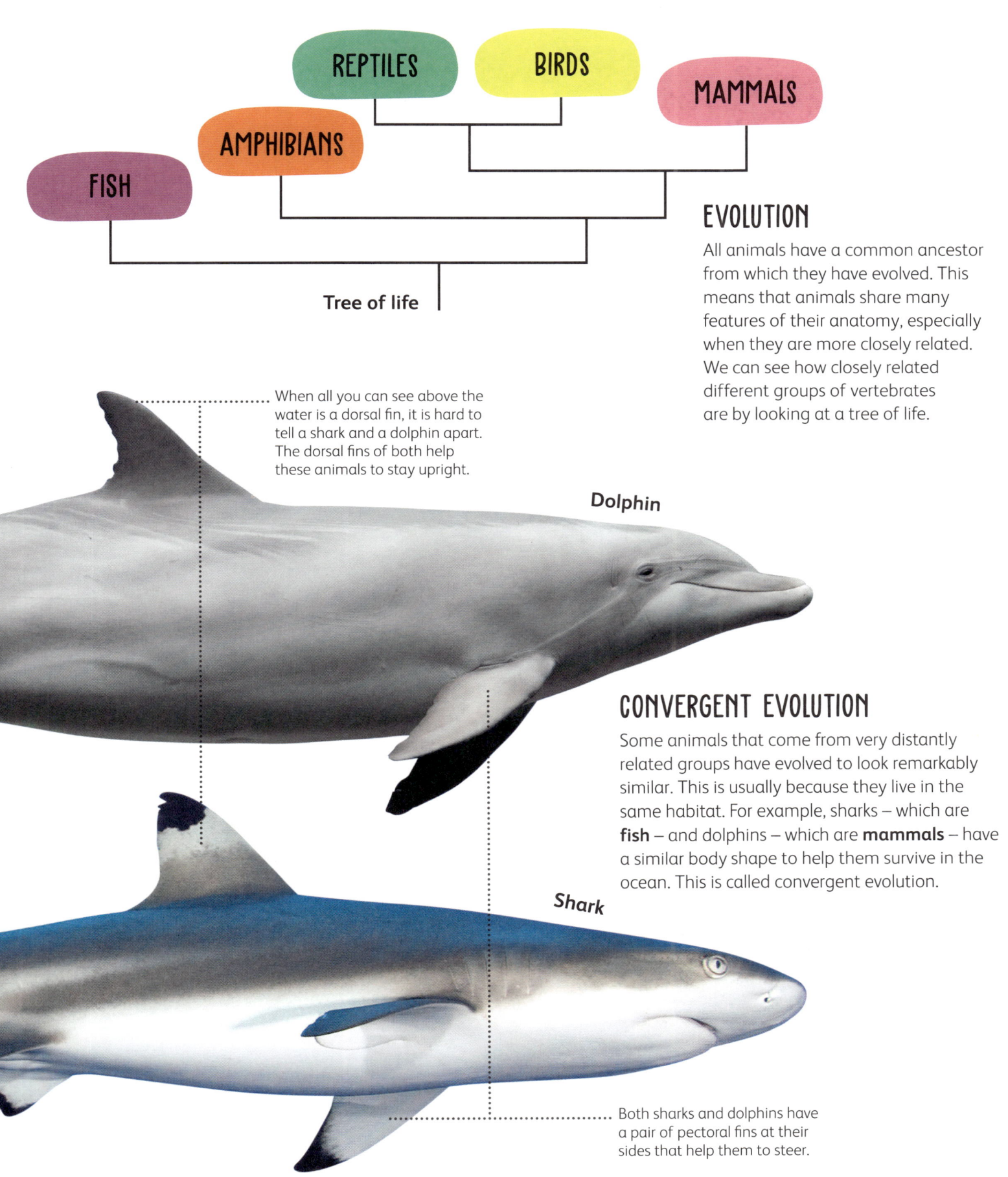

Tree of life

EVOLUTION

All animals have a common ancestor from which they have evolved. This means that animals share many features of their anatomy, especially when they are more closely related. We can see how closely related different groups of vertebrates are by looking at a tree of life.

When all you can see above the water is a dorsal fin, it is hard to tell a shark and a dolphin apart. The dorsal fins of both help these animals to stay upright.

Dolphin

CONVERGENT EVOLUTION

Some animals that come from very distantly related groups have evolved to look remarkably similar. This is usually because they live in the same habitat. For example, sharks – which are **fish** – and dolphins – which are **mammals** – have a similar body shape to help them survive in the ocean. This is called convergent evolution.

Shark

Both sharks and dolphins have a pair of pectoral fins at their sides that help them to steer.

SKELETONS

WHAT IS A SKELETON?

A **skeleton** is the sturdy frame that provides shape to the body of a **vertebrate** (animal with a **backbone**). Skeletons also protect vital **organs** like the **heart** and **brain**. They come in many different shapes and sizes, depending on what type of animal they belong to, where it lives, how it moves, and what it eats. Most vertebrates have skeletons made of **bones**.

WORKING TOGETHER

Bones don't work alone. They are joined to **muscles** and other bones by connections called **tendons** and **ligaments**. Together, bones and muscles must be strong enough to support the weight of the animal.

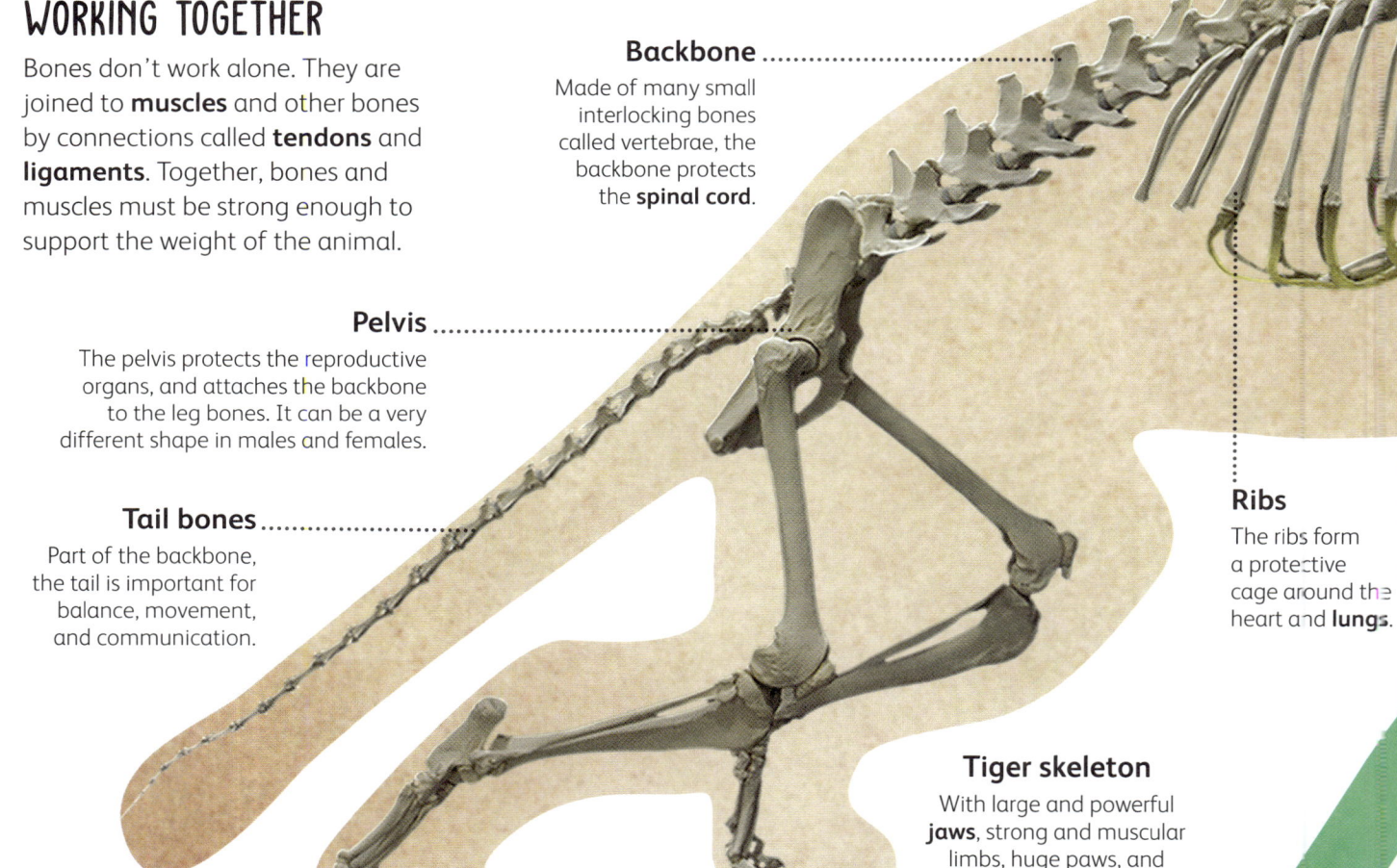

Backbone
Made of many small interlocking bones called vertebrae, the backbone protects the **spinal cord**.

Pelvis
The pelvis protects the reproductive organs, and attaches the backbone to the leg bones. It can be a very different shape in males and females.

Tail bones
Part of the backbone, the tail is important for balance, movement, and communication.

Ribs
The ribs form a protective cage around the heart and **lungs**.

Tiger skeleton
With large and powerful **jaws**, strong and muscular limbs, huge paws, and a bendy backbone, the tiger's skeleton is built to allow it to catch and bring down large **prey**.

Scapula
A flat, wide bone called a scapula connects the body to each forelimb.

Skull
Many bony plates join together around the brain, forming the **skull**.

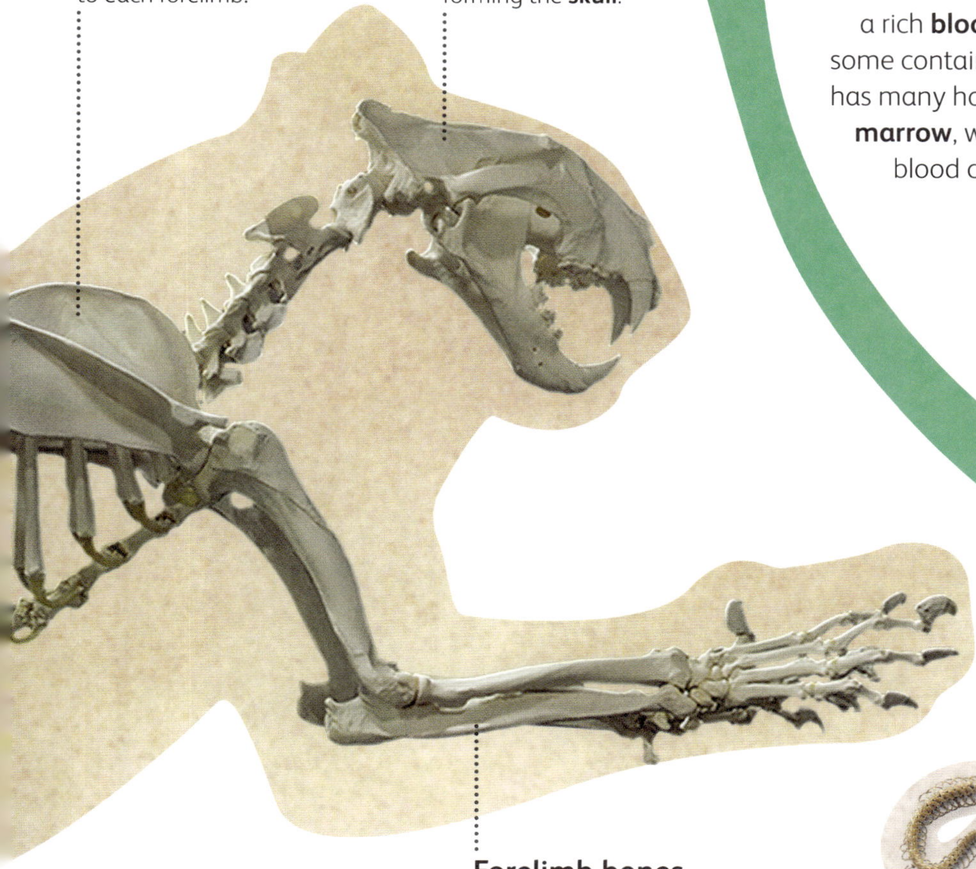

BONE STRUCTURE

We usually only see bones after an animal has died, when they are dusty and dry. But living bones are white and shiny with a rich **blood** supply. Inside, some contain spongy bone that has many holes filled with **bone marrow**, which is where red blood cells are made.

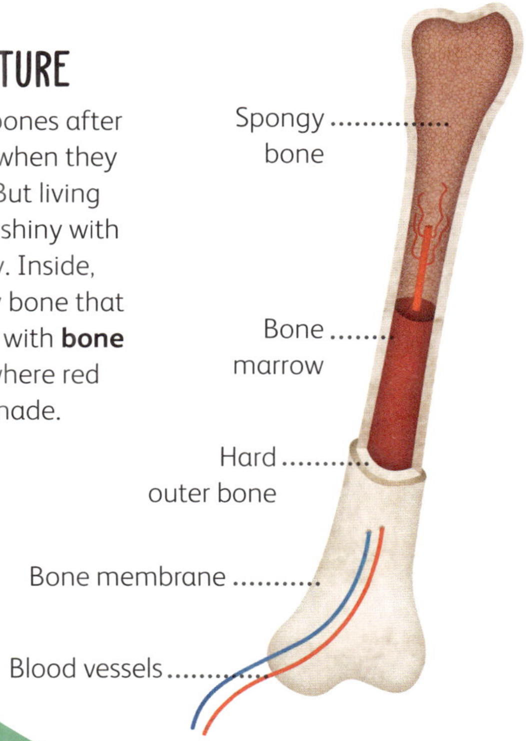

Spongy bone

Bone marrow

Hard outer bone

Bone membrane

Blood vessels

Forelimb bones
Thick, strong forelimb bones let the tiger cling on to prey when hunting.

JOINTS

Bones come together in areas called **joints**, which allow the skeleton to move in a variety of ways. The two bones in a joint are connected by tough cords, called ligaments. Inside many joints, the bones are cushioned by a firm but bendy **tissue** called **cartilage** and a fluid-filled capsule.

Bone
Membrane
Capsule
Cartilage
Fluid
Ligament

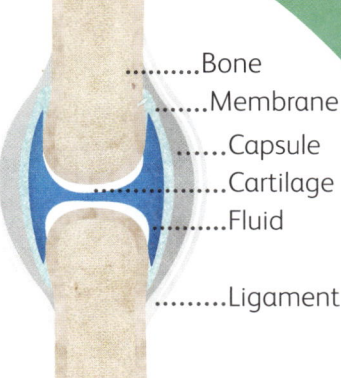

Snake skeleton
Snakes have extra long skeletons with hundreds of vertebrae, each with two ribs attached. Their long, limbless bodies are very flexible, and allow them to swallow huge meals.

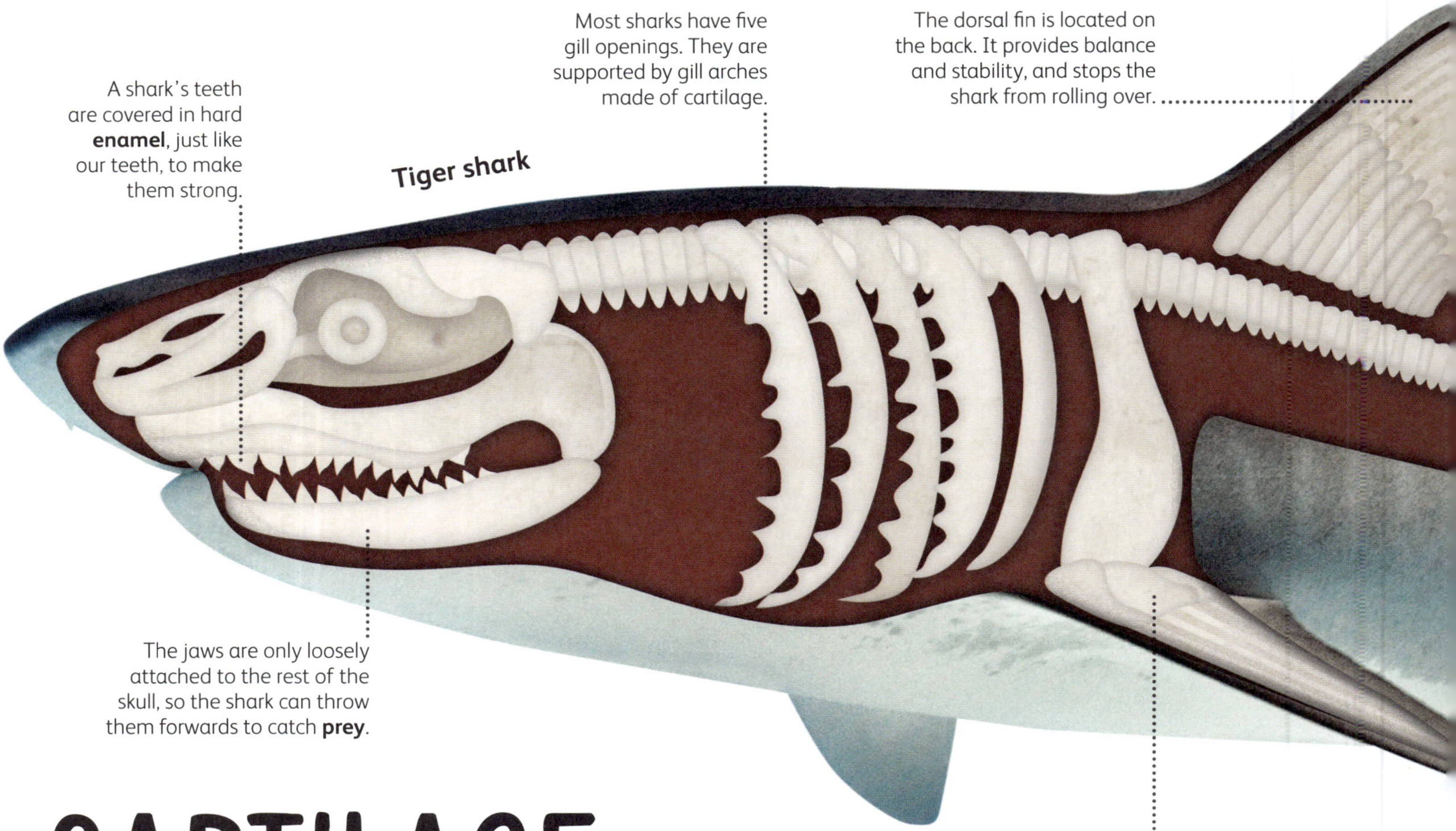

A shark's teeth are covered in hard **enamel**, just like our teeth, to make them strong.

Most sharks have five gill openings. They are supported by gill arches made of cartilage.

The dorsal fin is located on the back. It provides balance and stability, and stops the shark from rolling over.

Tiger shark

The jaws are only loosely attached to the rest of the skull, so the shark can throw them forwards to catch **prey**.

As well as providing balance, the pectoral fins are used to steer. There is one on each side of the shark's body.

CARTILAGE

Not all **vertebrates** have **skeletons** made from **bone**. **Cartilage** is a tough, rubbery, and bendy material that is softer than bone. In animals with bony skeletons it is found on the surfaces of **joints**, inside some **organs**, and in flexible body parts, such as the nose and ears. However, certain fish, including sharks, have entire skeletons made of cartilage. These animals belong to a group called **cartilaginous** fish.

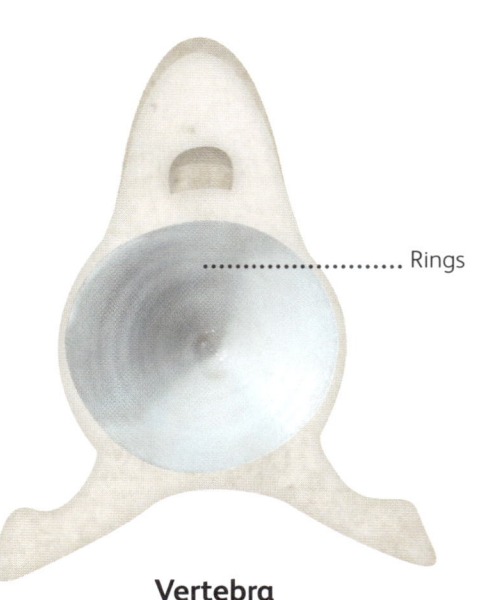

Rings

Vertebra

AGE OF A SHARK

In the circular centre of each vertebra of a shark is a pattern of rings, growing outwards like the circles inside a tree trunk. Just like a tree trunk, counting the number of rings can help tell you how old the individual is!

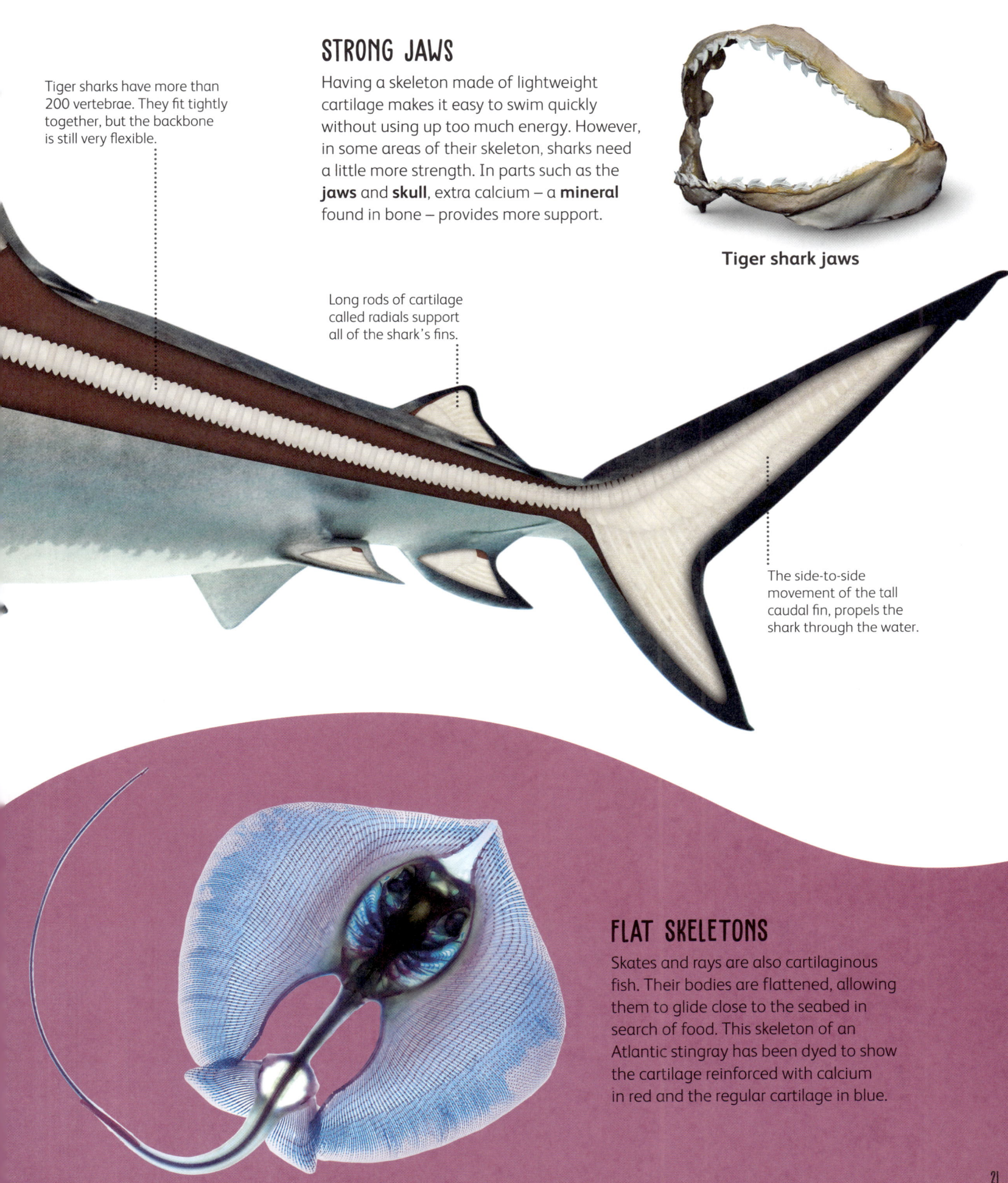

STRONG JAWS

Having a skeleton made of lightweight cartilage makes it easy to swim quickly without using up too much energy. However, in some areas of their skeleton, sharks need a little more strength. In parts such as the **jaws** and **skull**, extra calcium – a **mineral** found in bone – provides more support.

Tiger sharks have more than 200 vertebrae. They fit tightly together, but the backbone is still very flexible.

Tiger shark jaws

Long rods of cartilage called radials support all of the shark's fins.

The side-to-side movement of the tall caudal fin, propels the shark through the water.

FLAT SKELETONS

Skates and rays are also cartilaginous fish. Their bodies are flattened, allowing them to glide close to the seabed in search of food. This skeleton of an Atlantic stingray has been dyed to show the cartilage reinforced with calcium in red and the regular cartilage in blue.

STRETCHY JAW

A snake's lower jaw has two sides, which in life are connected at the front by a stretchy **ligament**. This means the snake can stretch its jaws wide to eat huge meals. Each of these sides can also move on its own, allowing the snake to "walk" its jaw along its **prey**.

Snake skull

ANTLERS

Antlers are extensions of the skull that are made of bone. They are shed and regrown every year. When they are growing, they are covered in a soft coating called velvet. Only members of the deer family, such as reindeer and moose, have antlers, and they use them to show off and protect themselves.

SKULL

An animal's head is filled with delicate **organs**, most importantly the **brain**. The head **bones**, more commonly known as the **skull**, create a tough case to protect them. The skull is split into two main parts: the mandible, which is the lower **jaw**, and the cranium, which is everything else. As well as offering protection, the skull provides a strong base for the **muscles** of the jaw to attach to, so an animal can bite and chew.

Turbinate bones

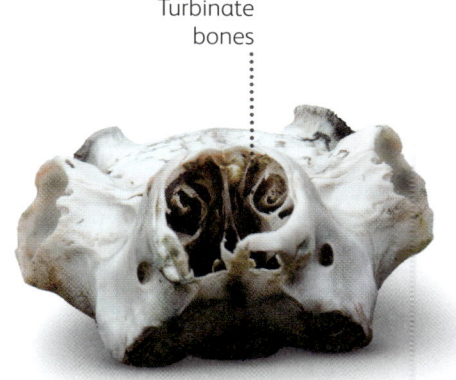

INSIDE THE NOSE

Inside the nose of **mammals** are delicate, thin bones that look like scrolls of paper. These are called turbinates and they are covered in a lining that moistens and warms air as it is breathed in. The complex folds of turbinates help to improve an animal's sense of smell.

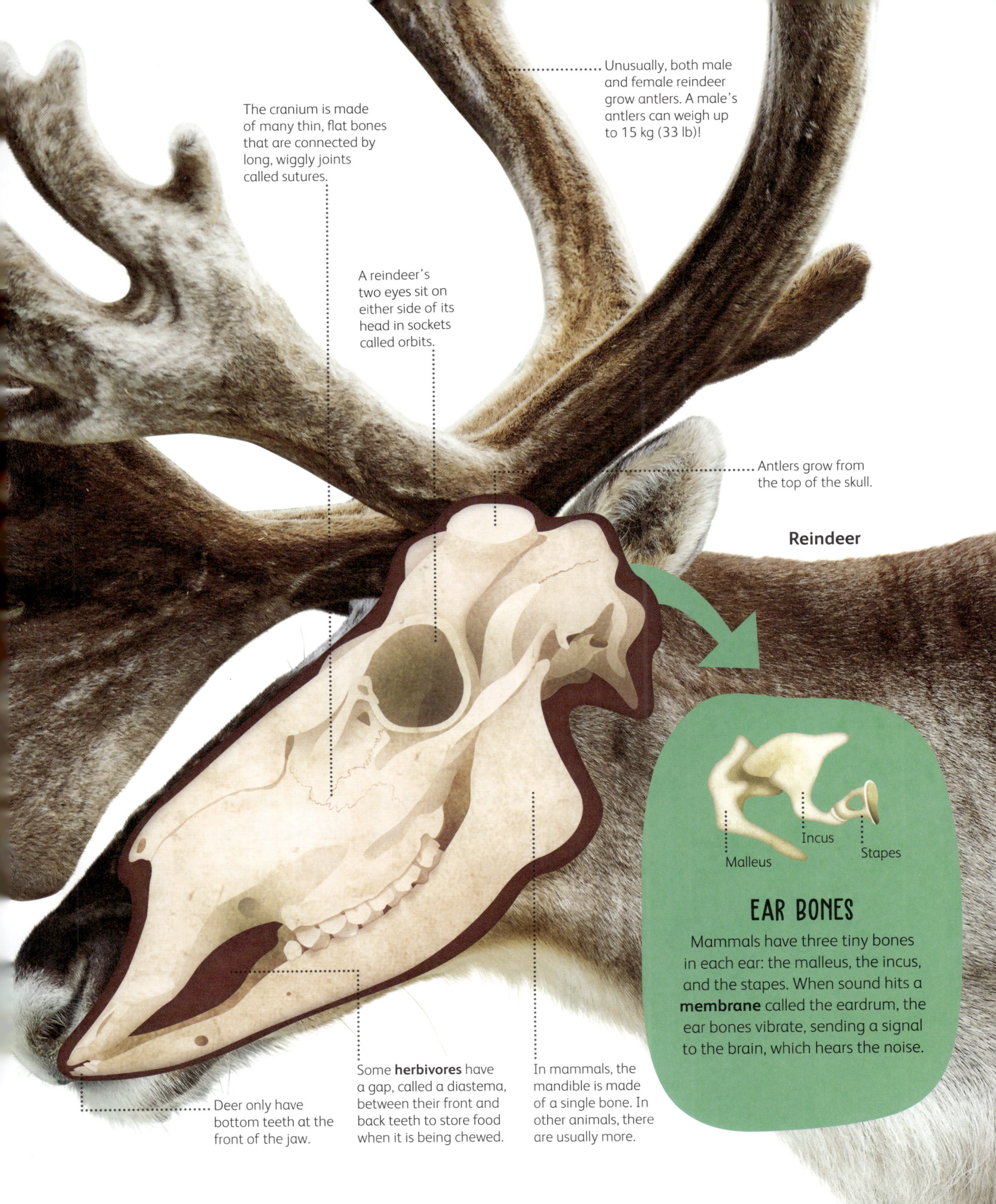

SKELETONS

FUSED BONES

Bones can be a variety of unusual shapes, but the shells of turtles and tortoises are some of the most extraordinary. Made of around fifty bones fused together, the domed upper shell, called a carapace, provides a tough cover for their internal **organs**. The lower shell, called a plastron, is flat, but just as hard as the carapace, and protects the soft underbelly.

The upper part of the shell is called the carapace.

A layer of scutes on top of the carapace gives greater protection to the shell.

Radiated tortoise

The pelvis joins the bones of the back legs to the shell.

Fused rib

Fused vertebra

The back legs are sandwiched between the top and bottom of the shell, restricting the tortoise's movements to a lumbering walk.

The bottom plate of the shell is called the plastron.

FUSED BACKBONE

Turtles and tortoises do not have a separate **backbone** inside their shell. Instead, their vertebrae and rib bones have fused together to form the carapace itself. This hard and heavy armour means that turtles and tortoises are much less flexible than other **reptiles**, but much more protected!

Gastralia

EXTRA RIBS

Though no other reptiles have quite such an impressive skeleton as a tortoise, crocodiles and alligators have an unusual feature of their own. They have gastralia, or "stomach ribs", which are extra rib-like bones in the skin of the belly that support the stomach. In fact, the plastron of tortoises probably **evolved** from gastralia.

The scapula, a bone in the shoulder, is long and slender. Unlike in other **vertebrates**, it sits inside the ribs.

Many tortoises can pull their head into their shell to avoid **predators**.

Scute   Bone of shell

SCUTES

The whole shell is covered in thin plates known as scutes. These scutes are made of **keratin**, the same material that makes up hair and nails. The scutes protect the bone underneath and can have beautiful patterns on them. As a tortoise or turtle grows, the old scutes are shed as bigger ones grow in their place.

25

SKELETONS

PNEUMATIC BONES

One of the greatest challenges faced by birds is that their **skeletons** must be light enough to let them fly, but strong enough to cope with the forces generated by flight. To keep their skeletons light, birds have a system of hollow bones, known as pneumatic **bones**, while to add strength, other parts of a bird's skeleton have fused together. Many animals have pneumatic bones in their **skull** to make them lighter.

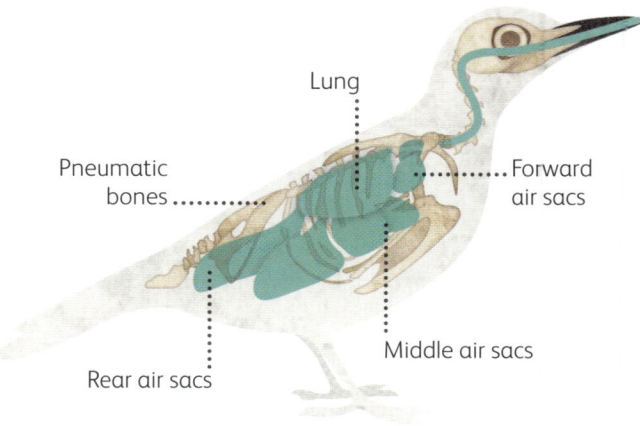

BREATHING BONES

Pneumatic bones are sometimes known as "breathing bones" because they connect to a bird's **lungs** and **air sacs** – balloons of air that help the bird breathe. Air moves from these air sacs into the pockets within the hollow bones to help the bird take in oxygen.

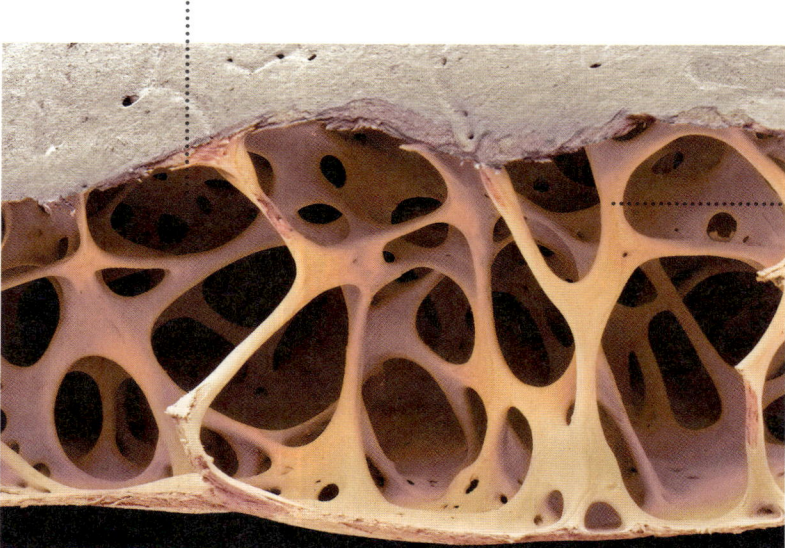

Pneumatic bones are filled with pockets of air.

Thin struts inside prevent the light pneumatic bones from breaking.

AIR POCKETS

Most of the bigger bones in a bird's skeleton, such as the keel and pelvis, are pneumatic. This means they contain large air pockets rather than **bone marrow**. The air pockets reduce the weight of the bones, but a network of thin struts inside keeps them strong. Many dinosaurs had pneumatic bones too.

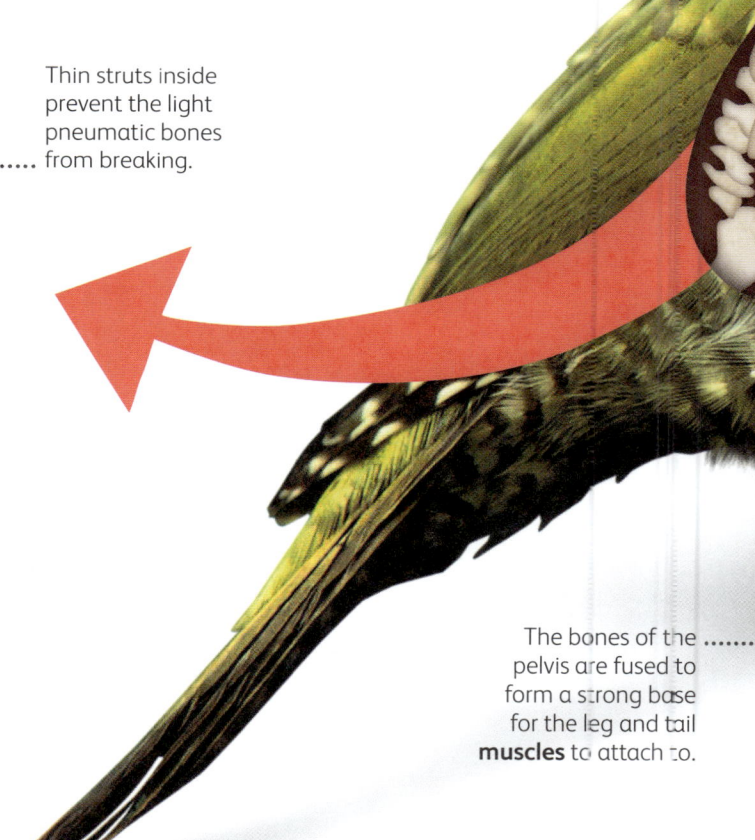

The bones of the pelvis are fused to form a strong base for the leg and tail **muscles** to attach to.

Green woodpecker

Many vertebrae, the ribs, and the keel are all pneumatic.

The large breastbone, called the keel, is where the flight muscles attach.

STRONG SKULL

Woodpeckers drum on wood for several reasons; to find food, to communicate with **mates**, and to excavate nests. It was thought that a special type of spongy bone found in the woodpecker's **skull** cushioned its **brain**, but it actually makes the skull stronger for hammering.

DENSE BONES

If penguins had pneumatic bones they would float to the top of the water! In fact, penguins have osteosclerotic bones, which are dense and solid, perfect for diving deep and helping the birds to stay underwater when hunting.

SKELETONS

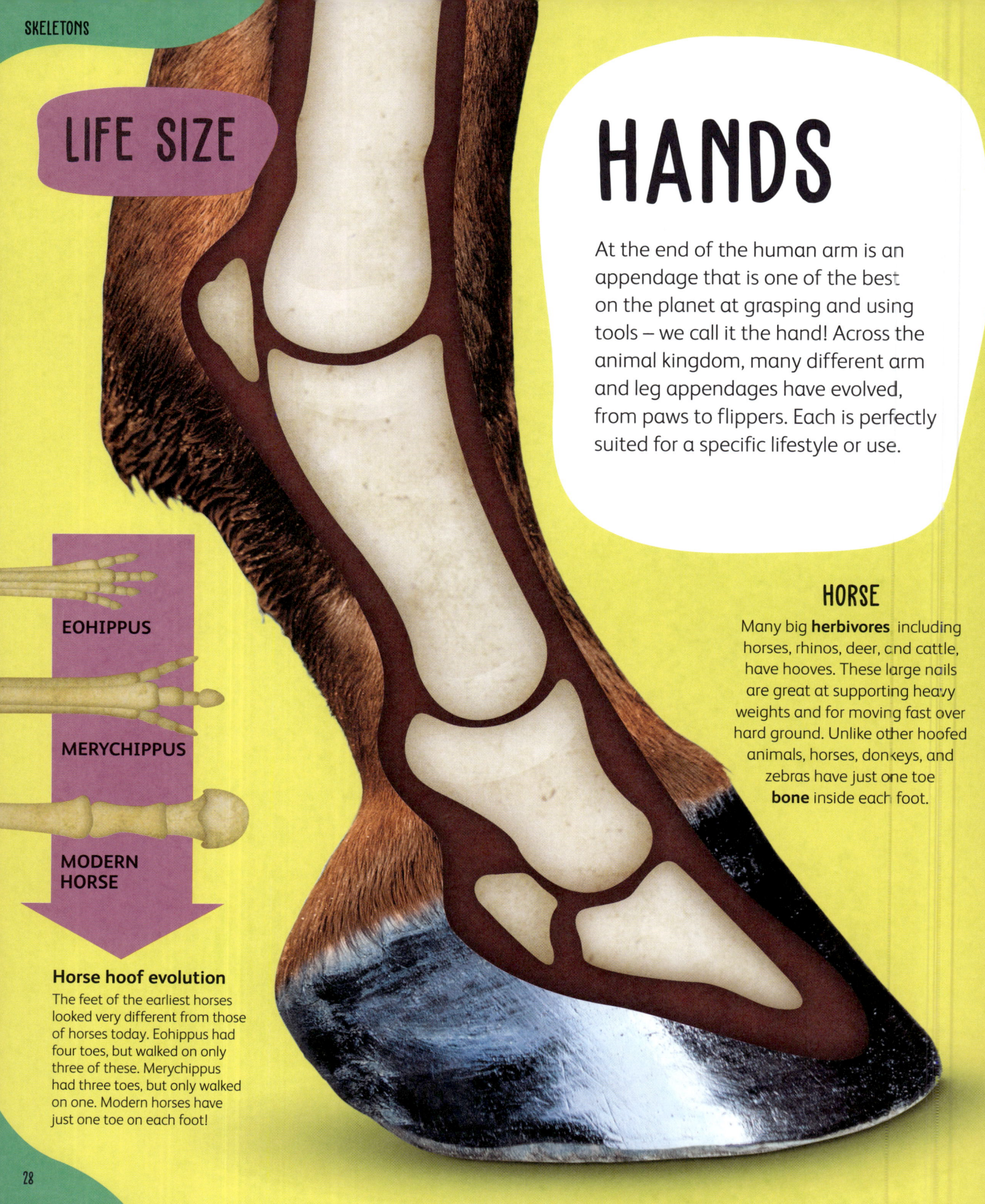

LIFE SIZE

HANDS

At the end of the human arm is an appendage that is one of the best on the planet at grasping and using tools – we call it the hand! Across the animal kingdom, many different arm and leg appendages have evolved, from paws to flippers. Each is perfectly suited for a specific lifestyle or use.

HORSE

Many big **herbivores** including horses, rhinos, deer, and cattle, have hooves. These large nails are great at supporting heavy weights and for moving fast over hard ground. Unlike other hoofed animals, horses, donkeys, and zebras have just one toe **bone** inside each foot.

EOHIPPUS

MERYCHIPPUS

MODERN HORSE

Horse hoof evolution
The feet of the earliest horses looked very different from those of horses today. Eohippus had four toes, but walked on only three of these. Merychippus had three toes, but only walked on one. Modern horses have just one toe on each foot!

WHALE

In place of hands, whales have flippers. These broad, flat appendages are used for steering underwater. There are separate finger bones hiding inside, but each flipper moves as one solid paddle.

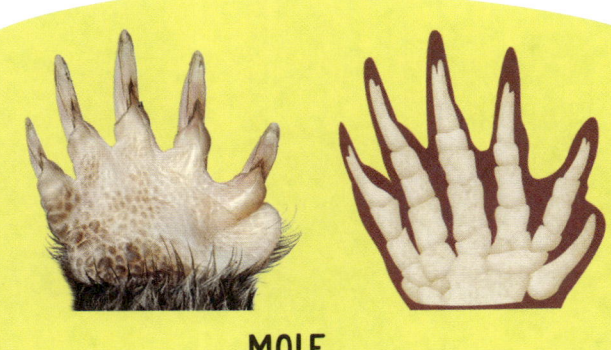

MOLE

Moles spend almost their entire lives in underground tunnels. They have paws shaped like shovels with strong claws that are perfect for digging. An extended wrist bone acts like an extra thumb, holding the paw rigid.

CHAMELEON

A chameleon's five fingers are split into two groups. Their Y-shaped hands are brilliant at clamping onto narrow branches, keeping them safe as they clamber through the trees.

SPIDER MONKEY

Spider monkeys have four long fingers but no visible thumb – although they still have thumb bones. This allows their hands to work like hooks as they swing through the trees.

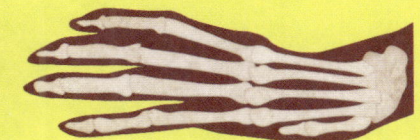

EAGLE

In eagles and other birds, the bones of the hand and arm are attached to lots of feathers, forming a wing. The wrist bones are fused to make the wing strong and there are only three small fingers.

SKELETONS

HAND AND FINGERS

Bat wings are made of **membranes** stretched between long finger **bones** and the bones in the arm, making them quite different from the wings of birds. In birds, the wing has fewer hand and finger bones but sturdy feathers give it strength. In bats, there are five fingers like other mammals, but they have become elongated to support the skin of the wing.

The thumb usually has a large claw. This is used for climbing, holding food, and fighting.

Bats can change the shape of their wings by moving their fingers.

Bat hands contain the same number of bones as human ones, but they are longer and thinner.

Bat bones contain fewer minerals than those of other mammals, making them very light.

The wing membrane, also known as the patagium, is a thin and hairless flap of skin, which pushes against the air during flight.

INSIDE A BAT

Only one **mammal** has truly mastered the art of flight – the bat. There are more than 1,400 types of bat in the world, meaning that bats make up around a fifth of all **species** of mammal! Despite their great numbers, we are often unfamiliar with these secretive animals. Under the surface, a set of remarkable **adaptations** allows these mysterious winged creatures to take to the night sky.

When the tensor tympani muscle contracts, it pulls on the malleus ear bone, which makes the ear less sensitive to loud noises.

When the stapedius muscle contracts, it pulls on the stapes ear bone, which makes the ear less sensitive to loud noises.

EAR BONES

Bats use **echolocation** to find their way in the dark – they make very loud, high-pitched sounds then listen for the returning echoes to build up a picture of their surroundings. So mammals don't damage their hearing when making loud sounds, **muscles** in the ear briefly pull on their ear bones to make them less sensitive. This is called the acoustic reflex.

Greater mouse-eared bat

BACKWARDS KNEES

A bat's back legs join its pelvis at a totally different angle to those of other mammals, resulting in knees that point backwards instead of forwards! This leg position makes it easier for bats to hang from roosts and to stretch out their wings in flight, but makes walking awkward.

A long spur of bone extending from the ankle, called a calcar, supports the membrane between the tail and the back legs.

Many bats have a tail membrane. This may be used to steer or to scoop **prey** towards the mouth.

MUSCLES

HORSE MUSCLES

There are around 700 muscles in a horse's body. Growing, powering, and repairing all of these muscles requires a lot of energy, which the horse gets from food and **oxygen**. Groups of muscles are found in layers that move different parts of the **skeleton**.

Neck muscles
Large neck muscles move the horse's head, but the head is held up by a strong **ligament** that connects the **skull** to the backbone.

Back muscles
Layers of muscles either side of the **backbone** work together to flex and bend it.

Hamstrings
Three powerful muscles in the back leg, known as the hamstrings, allow the horse to run, kick, and rear up.

Triceps
The triceps, a muscle which extends the elbow, is one of the few leg muscles a horse uses when standing still.

Abdominal muscles
The abdominal muscles are important for balance and support. As well as this, they help with breathing.

WHAT IS MUSCLE?

Muscle is a type of **tissue** made of many long **fibres**. Its job is to make things move. Every movement the body makes is controlled by muscles, from walking and jumping to the beating of the **heart** and the blinking of the eyes. To make the body move in a specific way, muscles shorten, or contract, to pull different body parts into different positions.

SKELETAL MUSCLE

Skeletal muscles are connected to **bones** by **tendons**. These are the muscles that allow animals to move around. They are controlled by signals from the **brain**, which are activated when an animal decides to move. Skeletal muscles are made of bundles of long **cells**, called fibres.

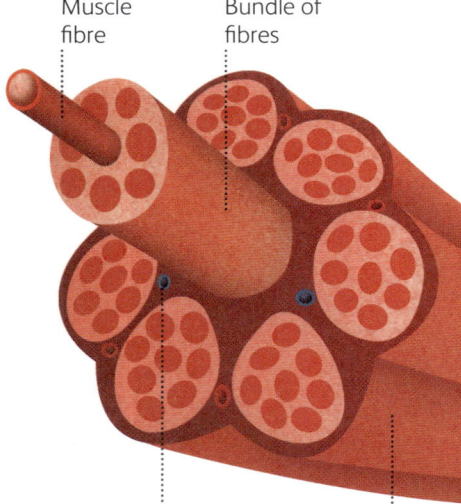

Muscle fibre
Bundle of fibres
Blood vessels
Muscle covering
Tendon

OTHER TYPES OF MUSCLE

Some types of muscle move automatically, without the animal having to think about it. They power important body processes, such as moving food through the digestive system. There are two varieties: cardiac muscle and smooth muscle.

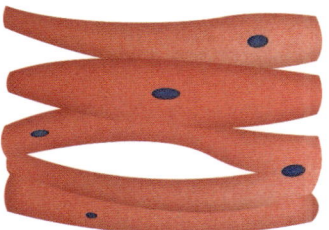

Cardiac muscle

Cardiac muscle is found in the walls of the heart. Coordinated contractions of cardiac muscles make the heart beat, which pumps **blood** around the body.

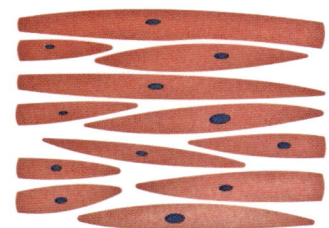

Smooth muscle

Smooth muscle is found in the walls of internal **organs**, such as the **intestines**, and in **blood vessels**. It squeezes these organs and vessels to help move what's inside them, like food or blood.

FAST AND SLOW MUSCLES

There are two types of skeletal muscle fibres, each of which power different types of movement. Fast twitch muscle fibres contract rapidly but soon get tired. Slow twitch muscle fibres contract more slowly but can continue working for a long time.

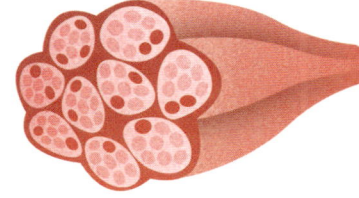

Fast twitch muscle

Muscles with more fast twitch fibres can work hard for a short time without oxygen. They are pale in colour. This type of muscle is used for sprinting.

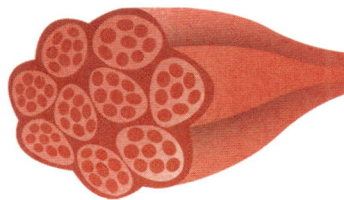

Slow twitch muscle

Muscles with more slow twitch fibres need lots of oxygen to work. They appear dark in colour. This type of muscle is used for walking long distances.

MUSCLES

Fiery-throated hummingbird

The shoulder **joint** of a hummingbird can be rotated 180 degrees, which makes the wing very flexible.

A **tendon** from the supracoracoideus muscle hooks over the shoulder, and attaches to the top of the humerus.

The pectoralis muscle is much larger in hummingbirds than in other types of bird so they can flap their wings faster.

The pectoralis attaches to the strong breastbone, called the keel.

HIGH ENERGY

Hummingbirds can fly at up to 97 kph (60 mph) and their wings can beat up to 80 times a second. That takes a lot of energy! To power those speedy wings, they spend most of their time hovering around flowers, using their long tongues to extract sugar-rich nectar.

PAIRED MUSCLES

Muscles are brilliant at pulling, but cannot push. When muscles become shorter and thicker, or contract, they can only pull a **bone** in one direction. So, to move in two directions, most bones are connected to two muscles with opposite actions – when one muscle contracts, the other relaxes. Muscles that work together like this are called **paired muscles**, such as the muscles that move a **bird's** wings up and down.

FLIGHT MUSCLES

Bird wings flap using two muscles. When the supracoracoideus muscle contracts, it pulls on a tendon that acts like a pulley to lift the humerus bone and move the wing up. This is called the upstroke. When the pectoralis muscle contracts, the wing moves down. This is the downstroke, and it generates most of the power.

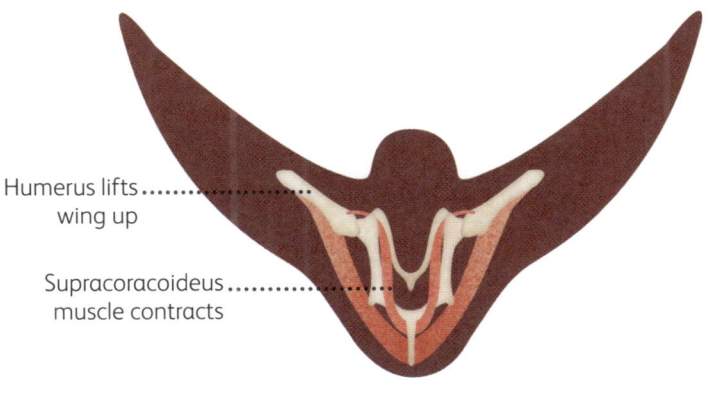

Humerus lifts wing up
Supracoracoideus muscle contracts

Upstroke

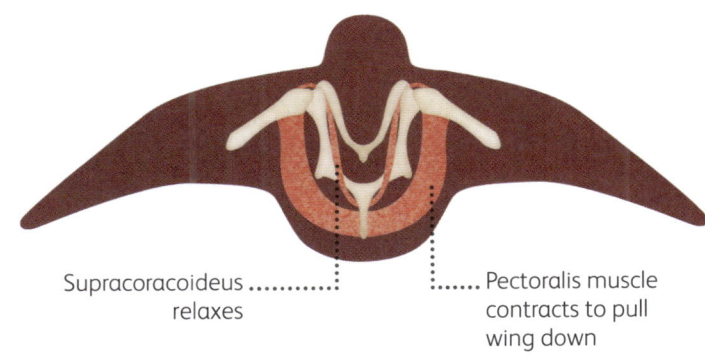

Supracoracoideus relaxes
Pectoralis muscle contracts to pull wing down

Downstroke

HOVERING

In order to hover in front of flowers, hummingbirds fly upright, while their wings move in a figure-of-eight pattern. This movement is only possible because of their flexible shoulder joint, and allows them to push themselves up on both the downstroke and the upstroke.

DOLPHIN SWIMMING

Paired muscles are responsible for the up-and-down movement of a dolphin's tailfin. **Fish** swim in a different way. Their paired muscles move their tails side-to-side to power them forwards.

TENDONS

TOE TENDONS

Toes are moved by long tendons that connect to muscles in the foot. Frogs have extremely long toes that are often joined together by flaps of skin called webs. The webbing turns the frog's feet into paddles, making them brilliant tools for propelling the frog through the water.

Tendons are like strong ropes that connect **muscles** to **bones**. When a muscle contracts it pulls on its tendon, which in turn pulls on the bone or **joint**, causing it to move. Tendons, which are made of a material called **collagen**, are extremely tough and can withstand enormous forces. They can also store up elastic energy, to help make certain movements extra powerful, such as when a frog jumps.

CALF CATAPULT

The secret weapons that allow a frog to jump great distances are its tendons. When the calf muscles in a frog's leg contract, they stretch a tendon in the ankle which stores up elastic energy. When the muscles relax, the tendon snaps back and the energy is released, catapulting the frog into the air.

One end of the tendon wraps around, and connects to, the heel bone or calcaneus.

QUICK ESCAPE

Being **amphibians**, frogs are **adapted** to life on land and in the water, and they are brilliant at both jumping and swimming. Jumping is a quick way to escape from **predators** and frogs are experts at it, sometimes reaching distances of more than 20 times their own body length in just one leap!

Common frog

The other end of the tendon connects to a muscle in the calf called the plantaris longus.

The frog has large calf muscles, which help power its jumps.

AUTOMATIC LOCKING

How do birds sleep in trees without falling off their perch? It's all to do with their tendons. In some birds, when the ankle and knee joints bend, tendons automatically pull their toes closed. The toes won't open again until the leg is straightened, meaning that the bird can safely relax its muscles.

MUSCLES

Western lowland gorilla

The temporalis muscle pulls the lower jaw up and moves it side to side.

In herbivores, the muscles of the jaw are made of slow twitch muscle, which doesn't tire as quickly as fast twitch muscle – useful for all that chewing!

The masseter is a muscle that closes the mouth. It is larger in herbivores because they need to chew more.

The buccinator muscle makes up part of the cheek.

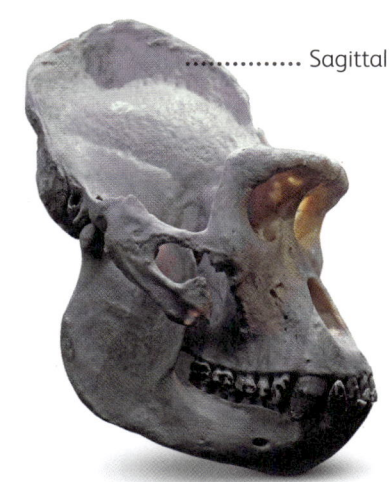

.... Sagittal crest

SAGITTAL CREST

The ridge at the top of a gorilla's **skull** is known as a sagittal crest. It provides a strong attachment point for the huge chewing muscles. Male gorillas have particularly large sagittal crests, which may have **evolved** to impress females.

NO CHEWING

Lions have very strong jaws, which they use to deliver killer bites to the necks of their **prey**. They have **adapted** to open their jaws wide and bite down hard, but this makes side-to-side movements difficult, so they usually tear their food into chunks instead of chewing it.

CHEWING

Animals chew their food to break it down into smaller pieces, which makes it easier for the body to take in the **nutrients** inside it. Chewing is particularly important for **herbivores**, which must work hard to extract the maximum energy from tough plants, and may feed for many hours a day. The **jaws** of plant-chomping animals often have large, powerful **muscles**, which keep their teeth moving.

EXPRESSIONS

Gorillas have a complex arrangement of muscles in their face that allows them to display a wide array of expressions. This allows them to show their emotions and communicate with other gorillas, helping them to make friends or avoid conflict.

MUSCLES

GRASPING BODY PARTS

Many animals can grab things with their hands, but some animals can use their feet or other body parts for grasping too. If a body part has **adapted** to wrap around and hold onto things, we call it **prehensile**. All kinds of body parts can be prehensile, from tails and noses to tongues and lips.

Grasping tongue
A giraffe's flexible tongue can easily strip the leaves off a tree branch. The tip of the tongue is full of **melanin**, which makes it black. This may be an adaptation to protect the tongue from sunburn.

SEAHORSE'S TAIL
Seahorses do not use their tail to swim, instead they use it to hold objects and anchor themselves to underwater plants and corals. Unusually, their tails are square-sided, which makes them extra good at grasping.

KINKAJOU'S TAIL
The kinkajou is a cat-sized mammal that lives in the canopy of tropical rainforests in Central and South America. Its long, prehensile tail allows the kinkajou to move nimbly through the trees and stops it from falling.

MANATEE'S LIPS
Manatees have a bristly upper lip which is split in half, creating two pads which can each move independently. They use their prehensile lip to grab aquatic plants and pass them into their mouth.

GIRAFFE'S TONGUE

The giraffe has a long, strong tongue, which it wraps around leaves and branches to tug them down from trees. It is thick and leathery to protect it from the thorns of the giraffe's favourite plant – the acacia tree.

LIFE SIZE

ELEPHANT'S TRUNK

As well as using them to grasp objects, elephants use their long noses – more commonly known as trunks – to drink, smell, and communicate. They can even use them as a snorkel when they are underwater!

SKINK'S TAIL

The Solomon Islands skink, a type of lizard, spends its entire life in the treetops, clambering from branch to branch. It is helped by its long, prehensile tail, which gives it its nickname of "monkey-tailed skink".

MUSCLES

INSIDE A CHAMELEON

Masters of **camouflage**, chameleons are brilliant at creeping about unseen. But their colouring is only one of the many extraordinary **adaptations** that make them such superb hunters. Thanks to an amazing set of **muscles**, chameleons can creep soundlessly through dense jungles, and use their unique eyes to pinpoint insect **prey** before catching them with a secret projectile weapon.

STICKY TRAP

To make sure their prey sticks to their tongue, chameleons produce a super sticky **mucus** that is 400 times thicker than human **saliva**. This honey-like glue traps large insects quickly before the muscular end of the tongue folds around the prey, so there is no escape!

TELESCOPIC TONGUE

The chameleon's sticky tongue can be fired out of its mouth to capture an insect. Inside the tongue is a **bone** surrounded by a sheath made of stretchy tubes that fit inside each other – like the rings of a collapsible telescope. Around the sheath is a muscle called the accelerator. When this muscle contracts, it pops open the tubes, shooting the tongue forward.

The tongue is anchored to the hyoid bone in the chameleon's neck.

At rest, the tubes of the sheath are stacked on top of each other.

There is a pointed bone inside the tongue.

When the mouth is closed, the accelerator muscle is relaxed.

The hyoglossus muscle pulls the tongue back into the mouth once the prey is captured, bunching up to fit inside.

The sheath catapults the tongue forwards with great force.

When the tongue fires, the hyoglossus muscle stretches to its full length.

A chameleon's tongue can stretch to twice its body length.

INDEPENDENT EYES

A chameleon can focus on two objects at the same time – one with each eye! This allows it to track two insects before deciding which one will make the tastiest lunch. When the target is locked in, the second eye swivels round to focus on it before the tongue is launched.

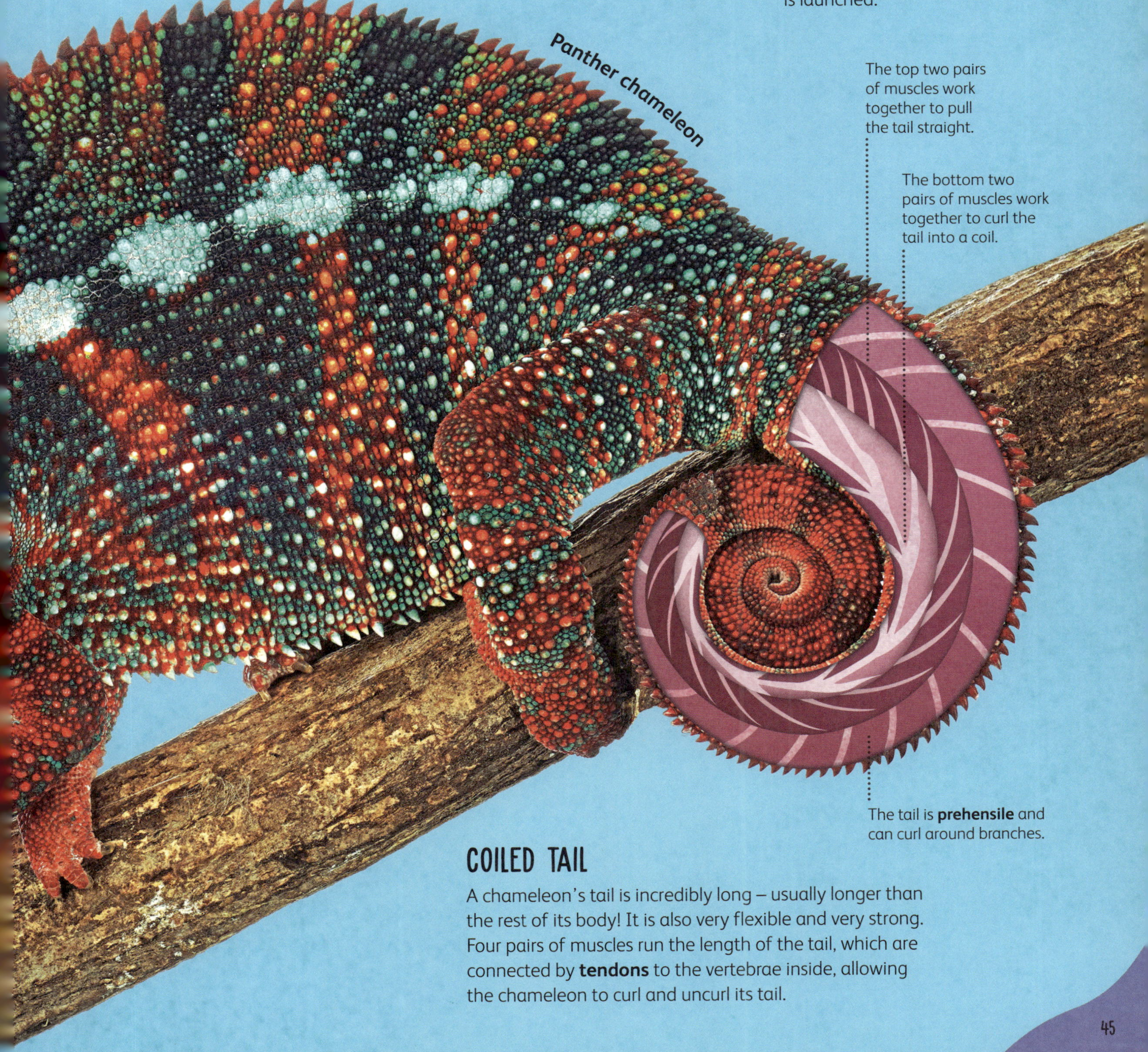

Panther chameleon

The top two pairs of muscles work together to pull the tail straight.

The bottom two pairs of muscles work together to curl the tail into a coil.

The tail is **prehensile** and can curl around branches.

COILED TAIL

A chameleon's tail is incredibly long – usually longer than the rest of its body! It is also very flexible and very strong. Four pairs of muscles run the length of the tail, which are connected by **tendons** to the vertebrae inside, allowing the chameleon to curl and uncurl its tail.

CIRCULATION

WHAT IS CIRCULATION?

Blood carries **oxygen** and **nutrients** to different parts of the body, including the **organs**, and collects their waste. This delivery system, or circulation, never stops. It is driven by a muscular pump called the **heart**, which pushes blood through special tubes known as **blood vessels**. When blood is full of oxygen, it is called **oxygenated**, and once it has delivered its oxygen to the organs, it is described as **deoxygenated**.

DOUBLE CIRCULATION

In many animals with **lungs**, the heart pumps blood around two different circuits; one carries blood to the lungs to pick up oxygen, and the other carries blood to the rest of the body to drop the oxygen off.

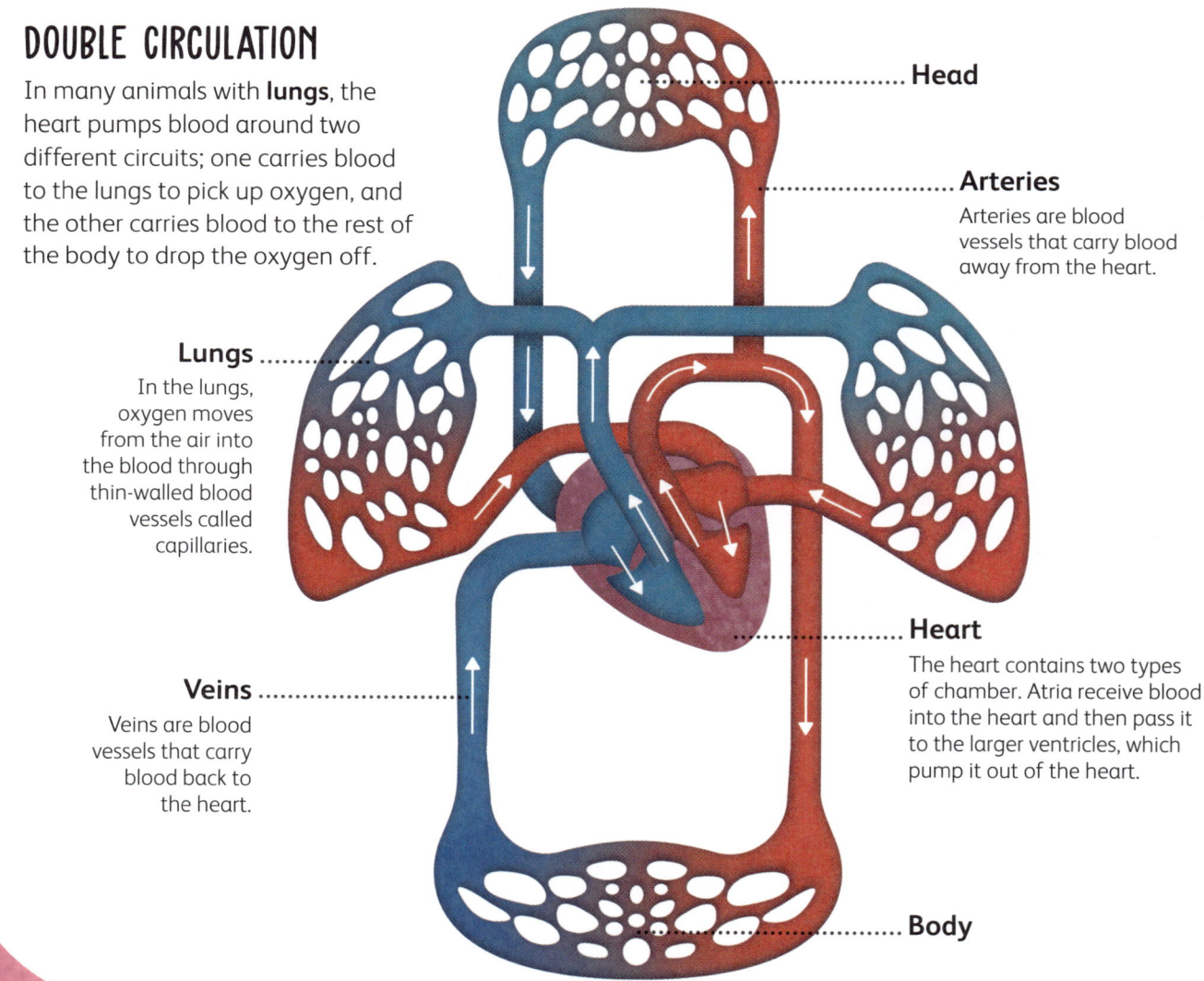

Head

Arteries
Arteries are blood vessels that carry blood away from the heart.

Lungs
In the lungs, oxygen moves from the air into the blood through thin-walled blood vessels called capillaries.

Heart
The heart contains two types of chamber. Atria receive blood into the heart and then pass it to the larger ventricles, which pump it out of the heart.

Veins
Veins are blood vessels that carry blood back to the heart.

Body

BLOOD VESSELS

There are three types of blood vessel, each with a different shape, depending on its function. Arteries are thick and stretchy to help pump blood to the body, veins are thin and floppy with **valves** to prevent blood flowing backwards, and tiny capillaries have very thin walls to let oxygen in and out of them.

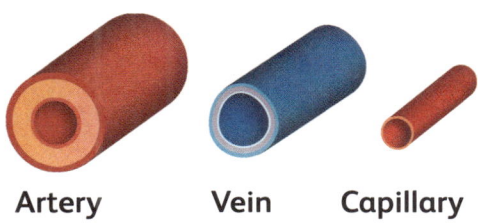

Artery **Vein** **Capillary**

HEARTS

The function of all hearts is to pump blood around the body. However, some animals have more sophisticated hearts than others, with more chambers to keep oxygenated and deoxygenated blood separate.

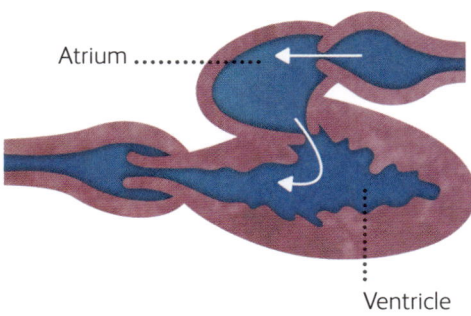

Fish

Fish hearts are S-shaped, with only one atrium and one ventricle. Most fish don't have lungs, so their circulation is on a single circuit. Blood goes straight from the **gills** to the rest of the body without returning to the heart first.

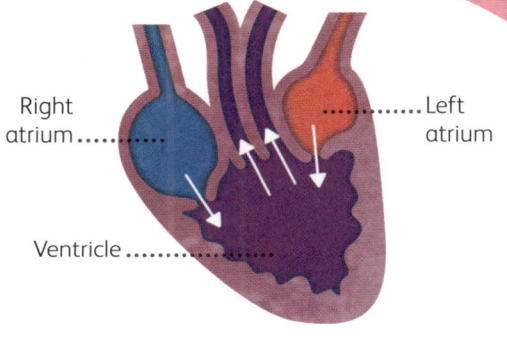

Amphibians

Amphibians have two atria and one shared ventricle. They have two different circuits, but the blood from each atrium can mix in the ventricle. The atria beat in a careful rhythm to stop the blood mixing too much.

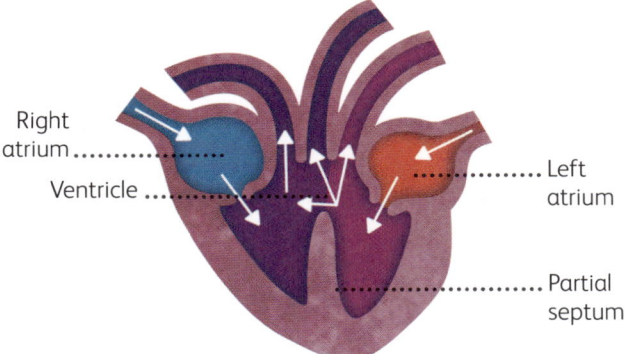

Reptiles

Like amphibians, reptiles have a three-chambered heart. However, it has a partial wall down the centre of the ventricle, called a septum, which means that blood from the two circuits is less likely to mix.

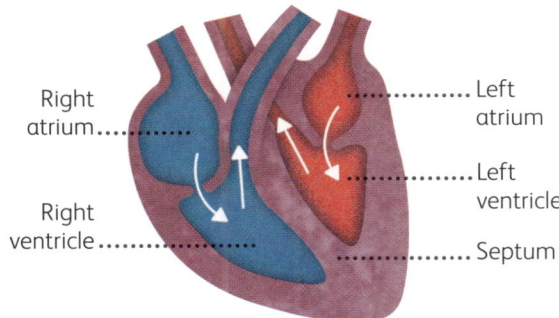

Birds and mammals

In birds and mammals the heart is totally divided into two sides, so oxygenated and deoxygenated blood don't ever mix. The right side sends blood to the lungs and the left side sends blood to the rest of the body.

CIRCULATION

HEART

The **heart** is one big pump. It sits in the chest, protected by a cage of **bones** – the ribs – and is connected to a system of **blood vessels** that lead all around the body. Each beat of the heart pushes **blood** around the vessels, keeping the other **organs** supplied with everything they need. Without a beating heart the entire body would soon shut down. Giant blue whales need one of the biggest hearts of any animal.

The beating of a blue whale heart is so loud it can be recorded by sonar equipment 32 km (20 miles) away.

Blue whale

HUGE SIZE
The blue whale is one of the largest animals to ever live – some grow to almost 30 m (98 ft) in length – so it makes sense that it has an enormous heart too. Its huge heart is around 1.5 m (5 ft) long and weighs about 180 kg (400 lb)!

HEART RATE
Usually, the smaller the animal, the faster its heart beats. The speed of a heartbeat is known as the **heart rate**. The heart of the tiny Etruscan shrew beats around 1,500 times a minute! In contrast, a human heart beats between 60 and 100 times a minute, and a blue whale's heart beats as little as 2 times a minute.

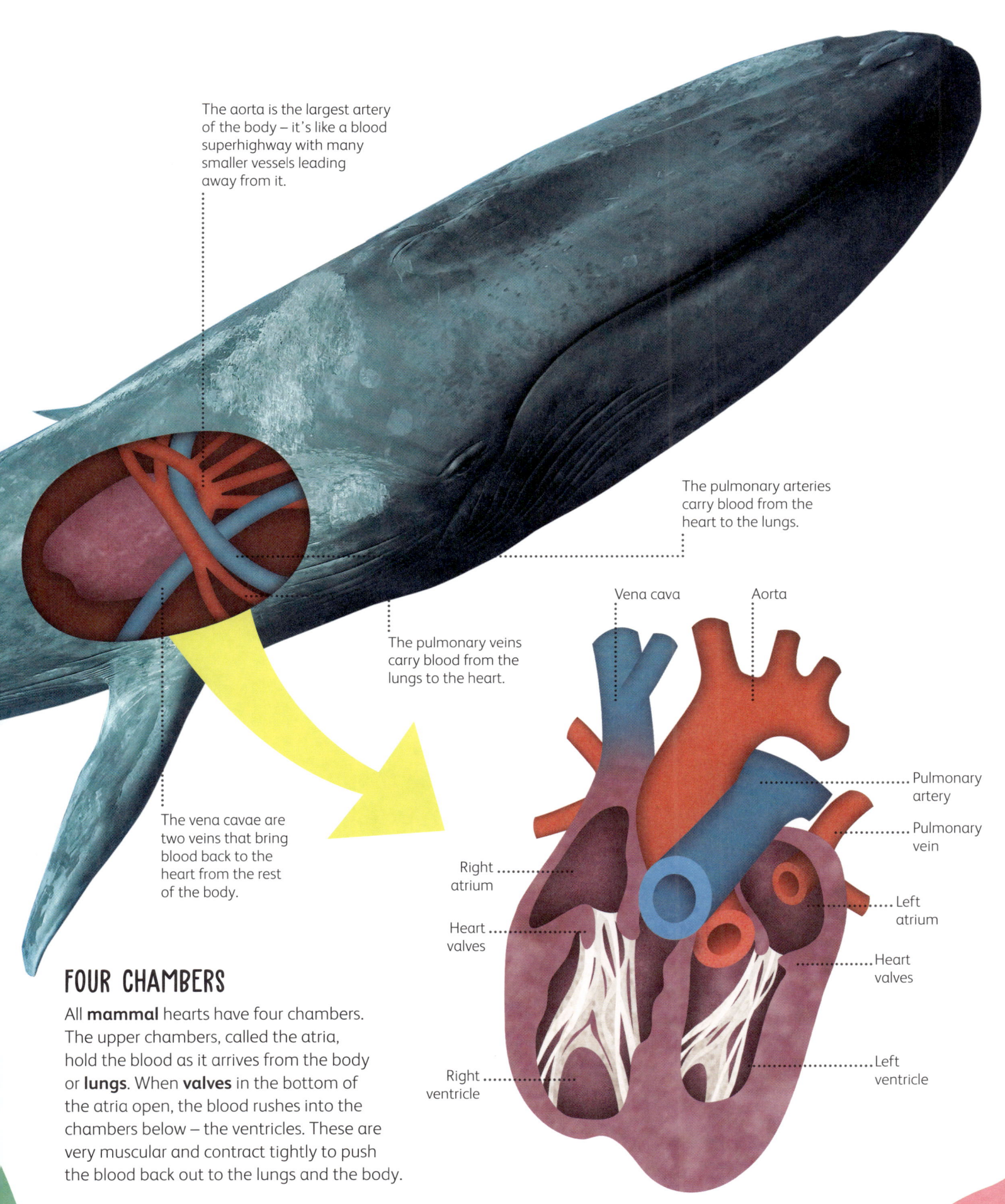

The aorta is the largest artery of the body – it's like a blood superhighway with many smaller vessels leading away from it.

The pulmonary arteries carry blood from the heart to the lungs.

The pulmonary veins carry blood from the lungs to the heart.

The vena cavae are two veins that bring blood back to the heart from the rest of the body.

Vena cava
Aorta
Pulmonary artery
Pulmonary vein
Right atrium
Left atrium
Heart valves
Heart valves
Right ventricle
Left ventricle

FOUR CHAMBERS

All **mammal** hearts have four chambers. The upper chambers, called the atria, hold the blood as it arrives from the body or **lungs**. When **valves** in the bottom of the atria open, the blood rushes into the chambers below – the ventricles. These are very muscular and contract tightly to push the blood back out to the lungs and the body.

SINGLE CIRCULATION

In **amphibians**, **reptiles**, **birds**, and **mammals**, **blood** travels through two different circuits — one to collect **oxygen** from the **lungs** and the other to deliver it to the body. In **fish**, though, after the blood is **oxygenated** in the **gills** it does not return to the **heart**. Instead, it flows directly around the body to the where it is needed. We call this a single circulatory system.

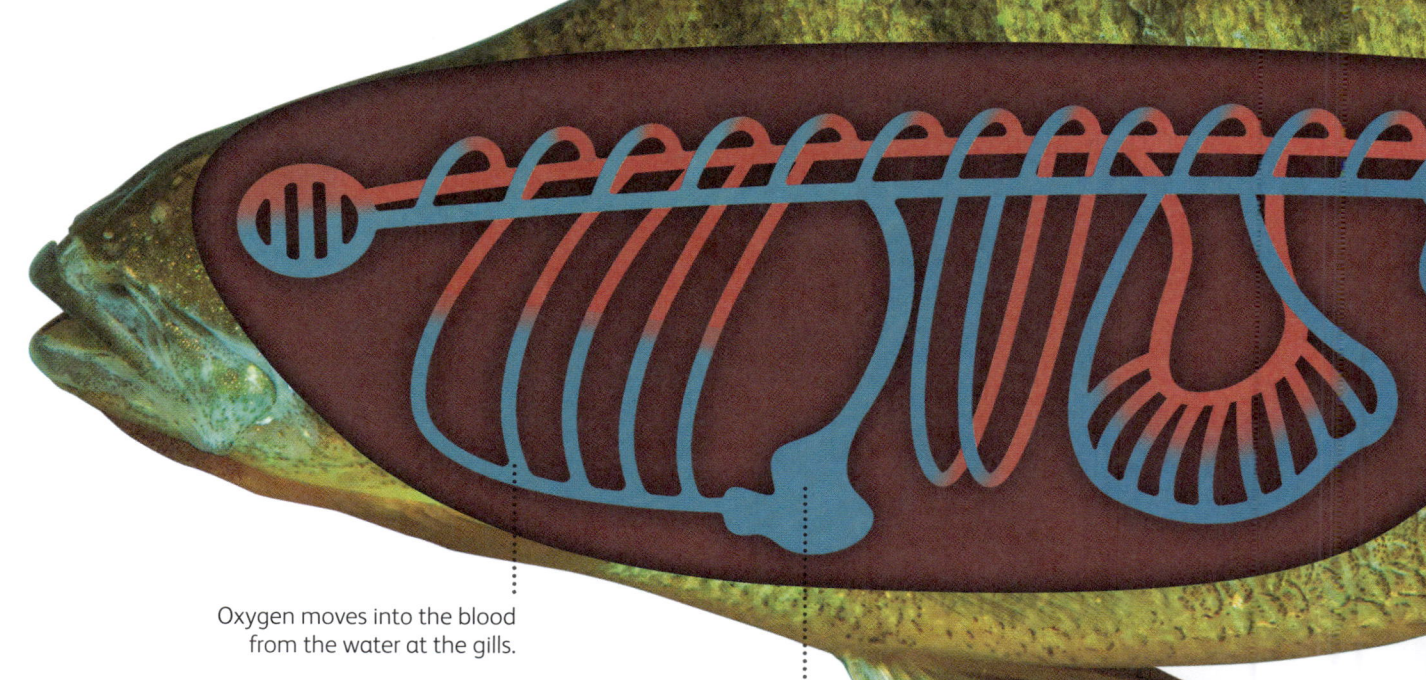

Redfin perch

Oxygen moves into the blood from the water at the gills.

Fish have a two-chambered heart with one atrium and one ventricle.

ABSORBING OXYGEN

In the gills, oxygen moves from the water into the bloodstream of the fish through the walls of tiny **blood vessels** called capillaries. Huge networks of capillaries lie just under the surface of the gills and have extremely thin walls, which allow gases to pass through them.

HAGFISH HEARTS

The hagfish has a very unusual circulatory system, which is unlike that of any other fish. As well as a central heart, it has three accessory hearts, which act as extra pumps. It also stores blood in large pools under its skin, called sinuses.

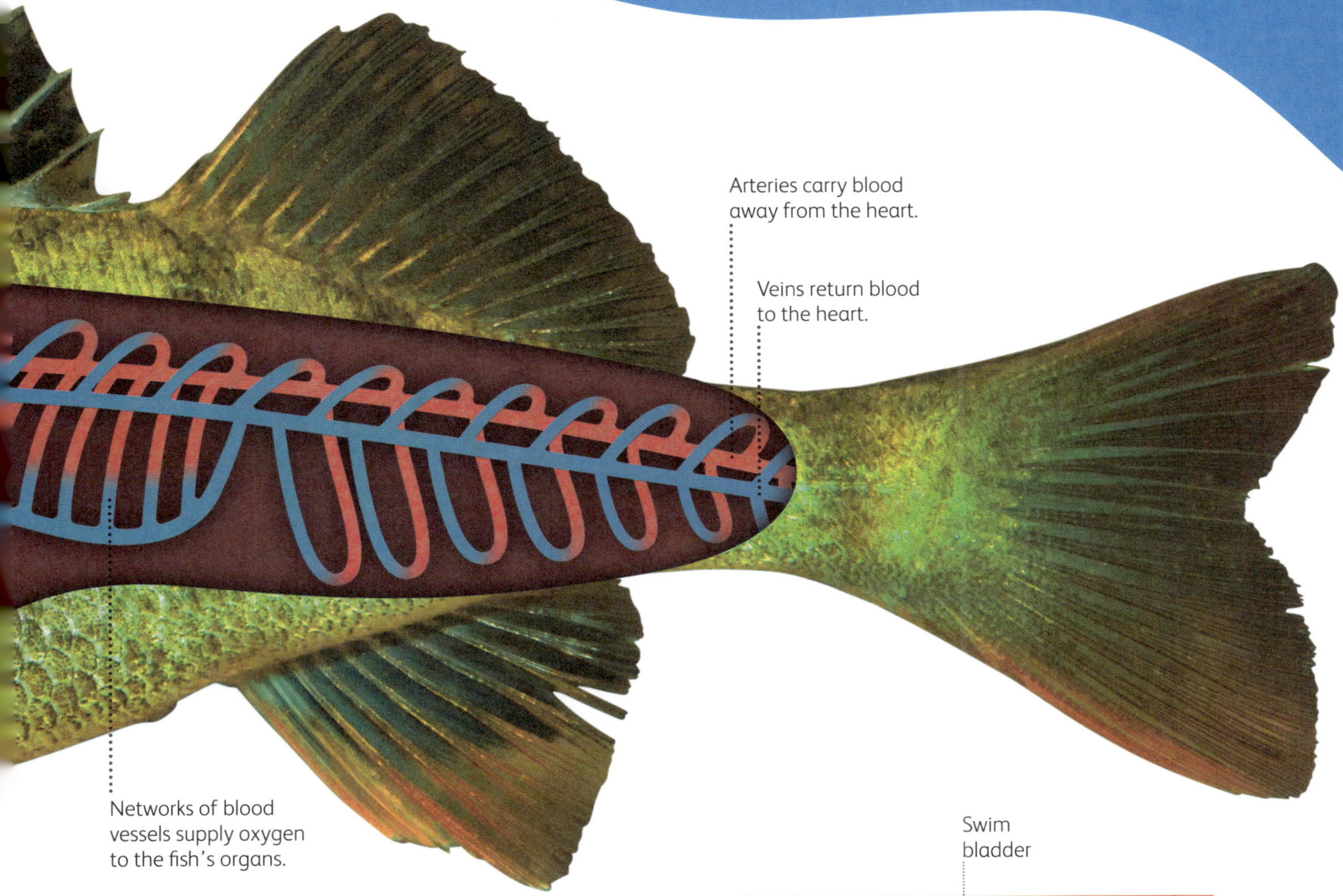

Arteries carry blood away from the heart.

Veins return blood to the heart.

Networks of blood vessels supply oxygen to the fish's organs.

SWIM BLADDER

The swim bladder is a gas-filled **organ** found in the abdomen of some fish. It helps fish to float in the water, and in some cases helps them make sounds. Gas enters and leaves the swim bladder from the blood through a network of capillaries, which are arranged in a dense, branching formation called a rete mirabile.

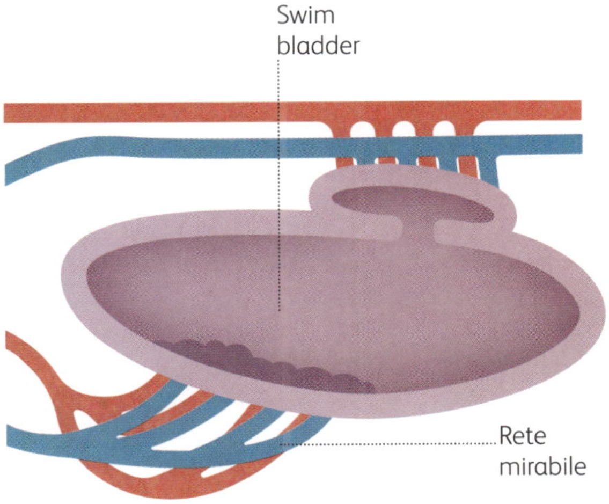

Swim bladder

Rete mirabile

CIRCULATION

BLOOD

So what exactly is **blood** and why is it so important? Blood is a fluid that contains all the things a body needs to grow, protect, and maintain itself. It is made of four main components: red blood cells, which carry **oxygen**; white blood cells, which defend the body against infections; platelets, which help to heal wounds; and plasma, which carries **nutrients** and waste products.

RED BLOOD CELLS

Most **mammals** have round red blood cells with a dip in the middle. This gives them a large surface area to help oxygen pass in and out. However, camels have small, oval-shaped red blood cells. This shape stops the cells getting stuck in tiny **blood vessels** when a camel goes for a long period without drinking and its blood becomes more concentrated.

Oval-shaped camel red blood cells move more easily through narrow blood vessels.

STAYING HYDRATED

Camels live in hot, dry areas like deserts, and can go for long periods without drinking. Despite amazing **adaptations** that help stop them becoming dehydrated, camels can lose up to 30 per cent of their body weight in water without becoming ill. This has a dramatic effect on their blood, which becomes much thicker.

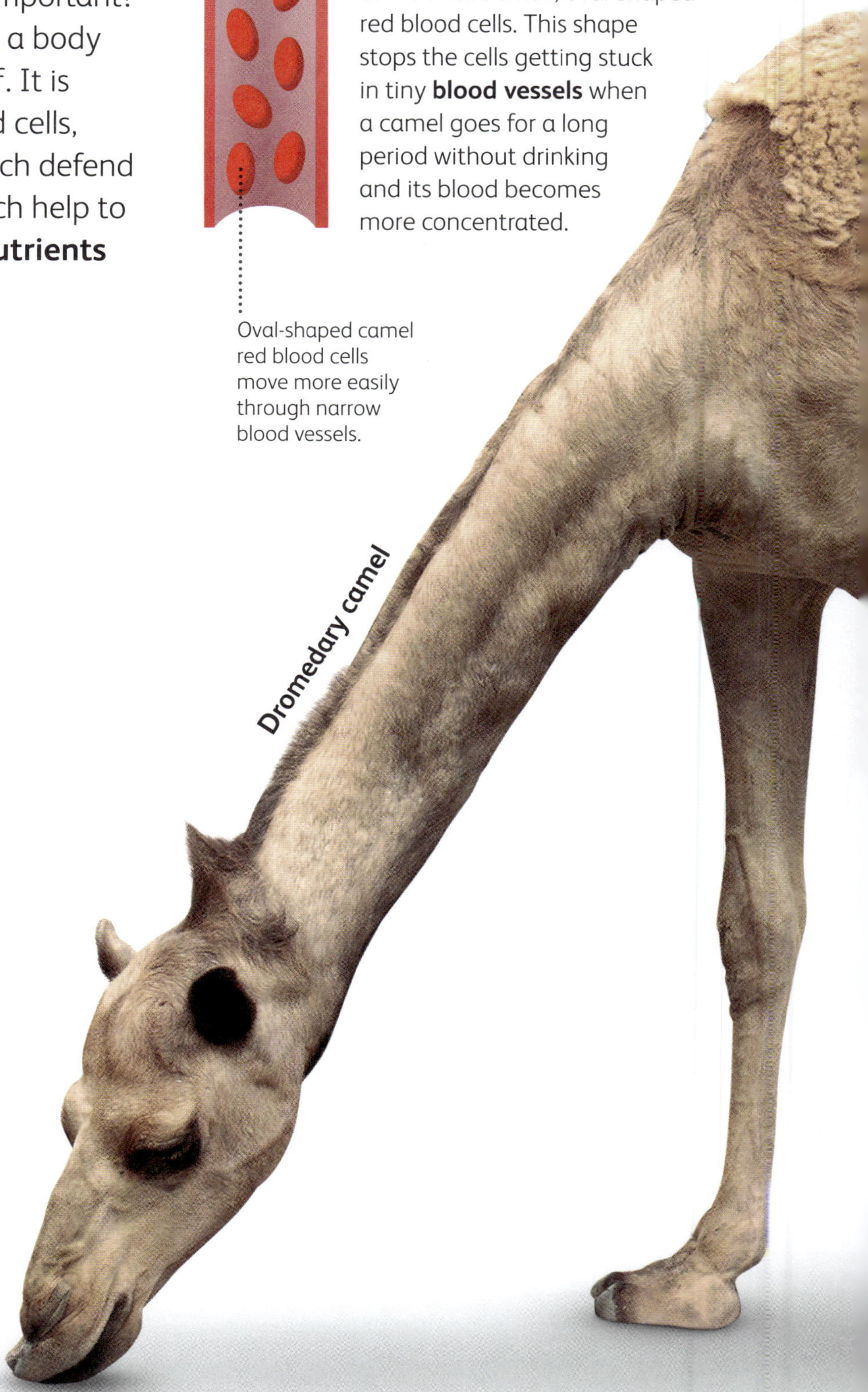

Dromedary camel

When a camel drinks lots of water, its oval-shaped red blood cells can more than double in size without bursting.

The camel's hump stores **fat**, which can be used as a source of energy when food and water are scarce.

The spleen is an **organ** that filters the blood, removing old or damaged red blood cells from circulation.

Camels reabsorb as much water as possible in their **intestines**, making their **urine** thick and syrupy, and their **faeces** hard and dry.

COLOURLESS BLOOD

The blood of almost all **vertebrates** is coloured red because of the red blood cells within it, but this is not the case with the Antarctic blackfin icefish. This unusual fish, which lives in the oxygen-rich, freezing cold waters around Antarctica, has no red blood cells. Instead, oxygen simply moves directly into its colourless blood from the water.

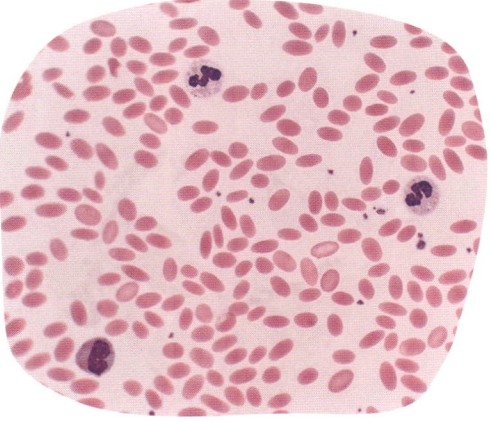

WHITE BLOOD CELLS

White blood cells protect the body against germs that cause illness, which are known as **pathogens**. There are different types of white blood cell, some of which kill pathogens directly, while others activate the **immune system** to prevent disease. In the image above, three large white blood cells – dyed dark purple – can be seen among the red blood cells – dyed pink.

CIRCULATION

TEMPERATURE

Every creature's body works best within a certain temperature range, but keeping within that range can be quite a task. Animals have **evolved** some smart ways of heating themselves up and cooling themselves down, allowing them to keep their body systems working properly, even in extreme **environments**.

COUNTERCURRENT HEAT EXCHANGE

The king penguin lives near the freezing Antarctic. To stop the penguin losing precious heat, the **blood vessels** carrying **blood** down its legs wrap closely around those carrying blood back from the feet. The warm blood from the body heats the cold blood as it returns from the feet, so the penguin's toes stay cold and its body stays warm! This system is called countercurrent heat exchange.

LIFE SIZE

BASKING

Some animals cannot create their own heat. To increase their temperature, they rely on external sources, such as sunlight or warm water. **Cold-blooded** lizards bask in the sun to raise their temperature.

VASODILATION

When blood is near the surface of the skin it is much easier for it to lose heat to the surrounding air. When an elephant is hot, blood vessels in its ears expand, or vasodilate, so that more blood can travel there to be cooled.

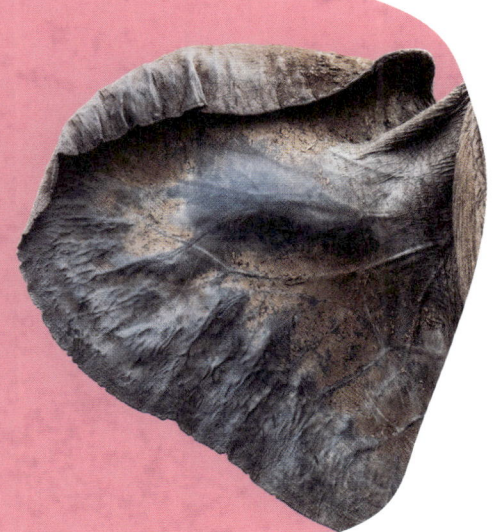

SWEATING

Many **mammals**, including horses, have special **glands** in their skin that release a watery liquid, called sweat, onto the surface of the skin through tiny holes, known as **pores**. When the sweat evaporates, it has a cooling effect on the body.

PANTING

Like all dogs, wolves breathe quickly, or pant, to push hot air out of the body and suck cool air back in. Keeping the mouth open also allows moisture to evaporate and cool blood in the mouth and tongue.

GULAR FLUTTERING

Birds don't have sweat glands, so they need to find other ways to stay cool – one way is called gular fluttering. A bit like panting, it's when birds rapidly flap the skin of their throats to force out hot air.

CIRCULATION

INSIDE A GIRAFFE

Giraffes tower over all other animals – commonly reaching heights of more than 5 m (16 ft). While being so tall is handy for snacking on leaves from high branches, it also presents several problems. Pumping **blood** all the way up that long neck requires a great deal of effort, and being so large makes it hard for giraffes to lose heat. Clever **adaptations** allow these gentle giants to battle gravity and stay cool.

The carotid arteries carry blood from the heart to the brain.

Giraffe

When the giraffe's head is lowered, valves in the jugular veins close to prevent blood flowing backwards.

When the giraffe's head is raised, normal flow resumes in the jugular veins.

The giraffe's jugular veins carry blood from the brain to the heart. They are around 2.5 cm (1 in) wide.

VEIN VALVES

When a giraffe needs to drink, it must lower its head all the way to the floor. Even with its legs wide apart, the giraffe's head is more than a metre below its **heart**. A rush of blood to the giraffe's **brain** would be damaging, but **valves** in the **veins** of the neck prevent blood flowing backwards to the brain.

MUSCULAR HEART

For the weight of its body, a giraffe's heart is about average in size, but it does have one striking difference – the walls of its ventricles are extremely thick. This gives it extra pumping power and creates the high blood pressure necessary to force blood uphill to the giraffe's brain.

A network of capillaries is found under the giraffe's spots.

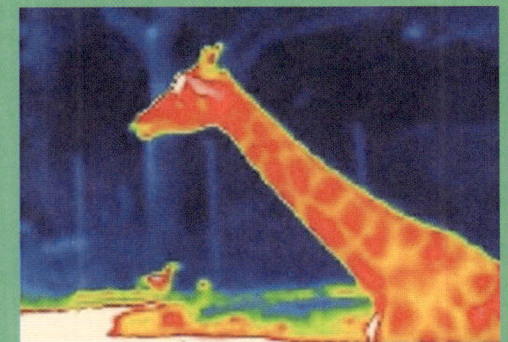

HOT SPOTS

In the baking savannahs of Africa where giraffes live, their pattern of spots not only provides excellent camouflage, but also keeps them cool. Beneath each dark spot is a network of **capillaries** that radiates heat away from the skin. You can see this in the **infrared** picture above, which shows hotter areas in red and cooler areas in blue.

A giraffe's heart can weigh more than 7 kg (15 lb). That's about 28 times more than a human heart.

The arteries in a giraffe's lower leg are narrower than in its upper leg.

A giraffe's legs can be 2 m (6½ ft) long.

For its head to reach the ground, a giraffe must nearly do the splits.

COMPRESSION SOCKS

Gravity wants to pull all of the giraffe's blood down into its feet but the tight **connective tissue** around its lower legs acts like super-elastic socks, preventing too much blood from entering and pooling there. Blood flow to the feet is also slowed down by **arteries** that narrow at the giraffe's knees.

59

RESPIRATION

Outside gills
Unlike the hidden gills of fish, amphibian gills are large, feathery, and on the outside of their body.

Capillaries
The gills are packed with blood-filled **capillaries**, which often makes them look red.

Branches
Gills have branches or folds to make their surface area larger, allowing them to extract the most oxygen they can.

Water flow
Water must flow constantly over an animal's gills in order for it to get enough oxygen.

GILLS

Many aquatic animals get their oxygen from water, which contains much less oxygen than air. In **fish** and some **amphibians**, such as the axolotl, oxygen is moved from water to the blood by special **organs** called **gills**.

WHAT IS RESPIRATION?

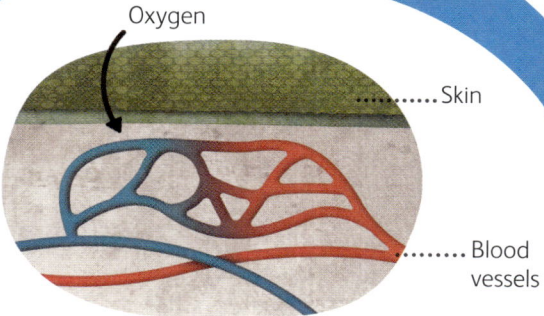

Oxygen
Skin
Blood vessels

OXYGEN THROUGH SKIN

In some animals, particularly amphibians, oxygen can move directly across the skin. For this to happen, the skin must be very thin, moist, and filled with a rich blood supply.

Animal bodies are constantly working — pumping **blood**, digesting food, getting rid of waste — and all of those actions need energy! Most animals get this energy from a process called **respiration**. In this process, **oxygen**, a gas found in the atmosphere, is combined with the **sugars** in food to release energy. Animals extract the oxygen they need for respiration in a number of different ways.

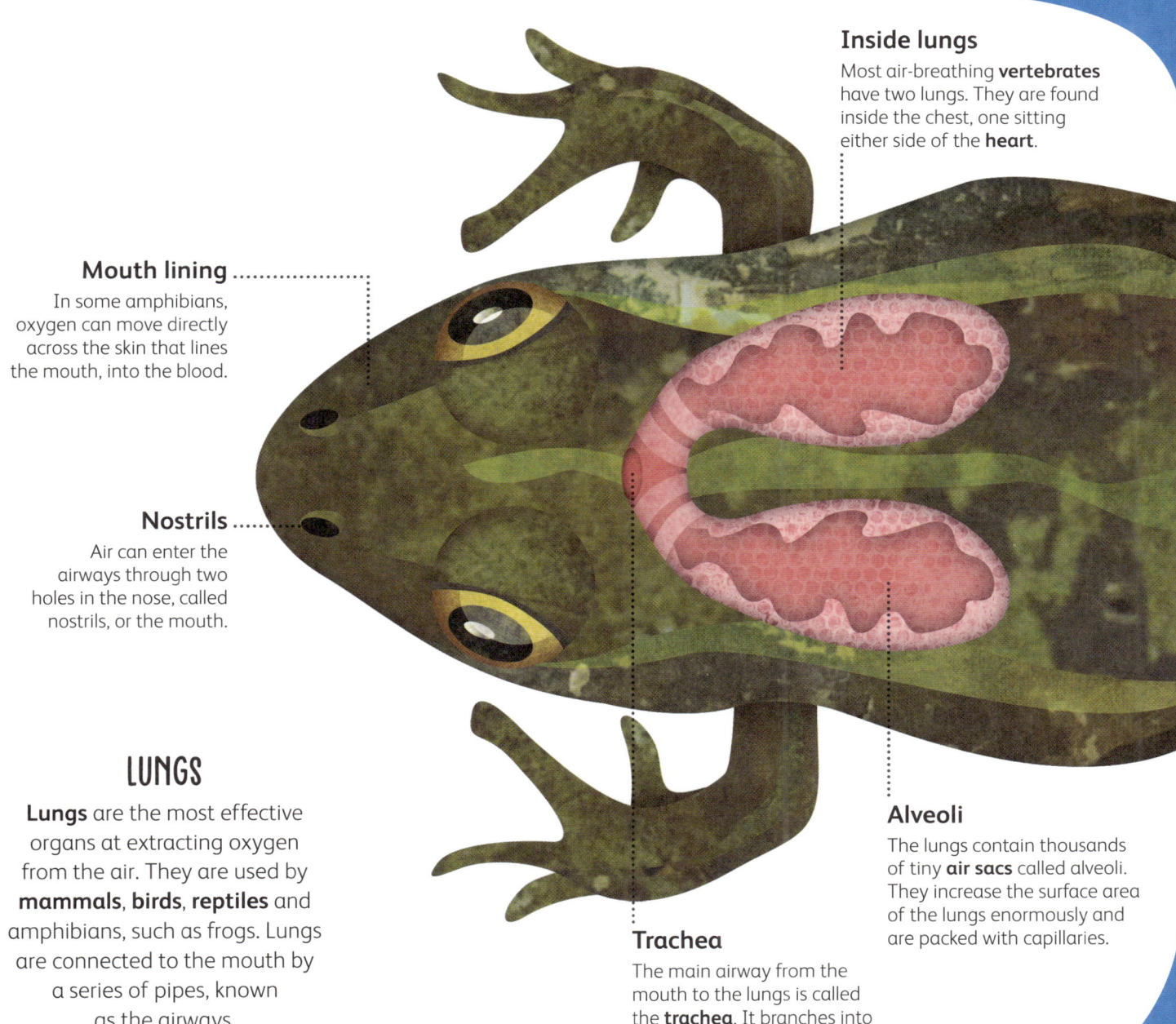

Inside lungs
Most air-breathing **vertebrates** have two lungs. They are found inside the chest, one sitting either side of the **heart**.

Mouth lining
In some amphibians, oxygen can move directly across the skin that lines the mouth, into the blood.

Nostrils
Air can enter the airways through two holes in the nose, called nostrils, or the mouth.

LUNGS
Lungs are the most effective organs at extracting oxygen from the air. They are used by **mammals**, **birds**, **reptiles** and amphibians, such as frogs. Lungs are connected to the mouth by a series of pipes, known as the airways.

Alveoli
The lungs contain thousands of tiny **air sacs** called alveoli. They increase the surface area of the lungs enormously and are packed with capillaries.

Trachea
The main airway from the mouth to the lungs is called the **trachea**. It branches into smaller tubes called bronchi.

GLUCOSE + OXYGEN → RESPIRATION → CARBON DIOXIDE + WATER + ENERGY

Respiration
Respiration is a **chemical reaction** between a sugar called **glucose** and oxygen. The reaction happens inside an animal's **cells** to produce energy. As well as releasing energy, respiration produces **carbon dioxide** and water as waste materials.

RESPIRATION

PROTECTIVE COVER

In some fish, the gills are covered by a bony plate called an operculum, which protects them from damage. The opening and closing of the operculum can help pump water over the gills, meaning these fish don't have to swim continuously in order to get enough oxygen.

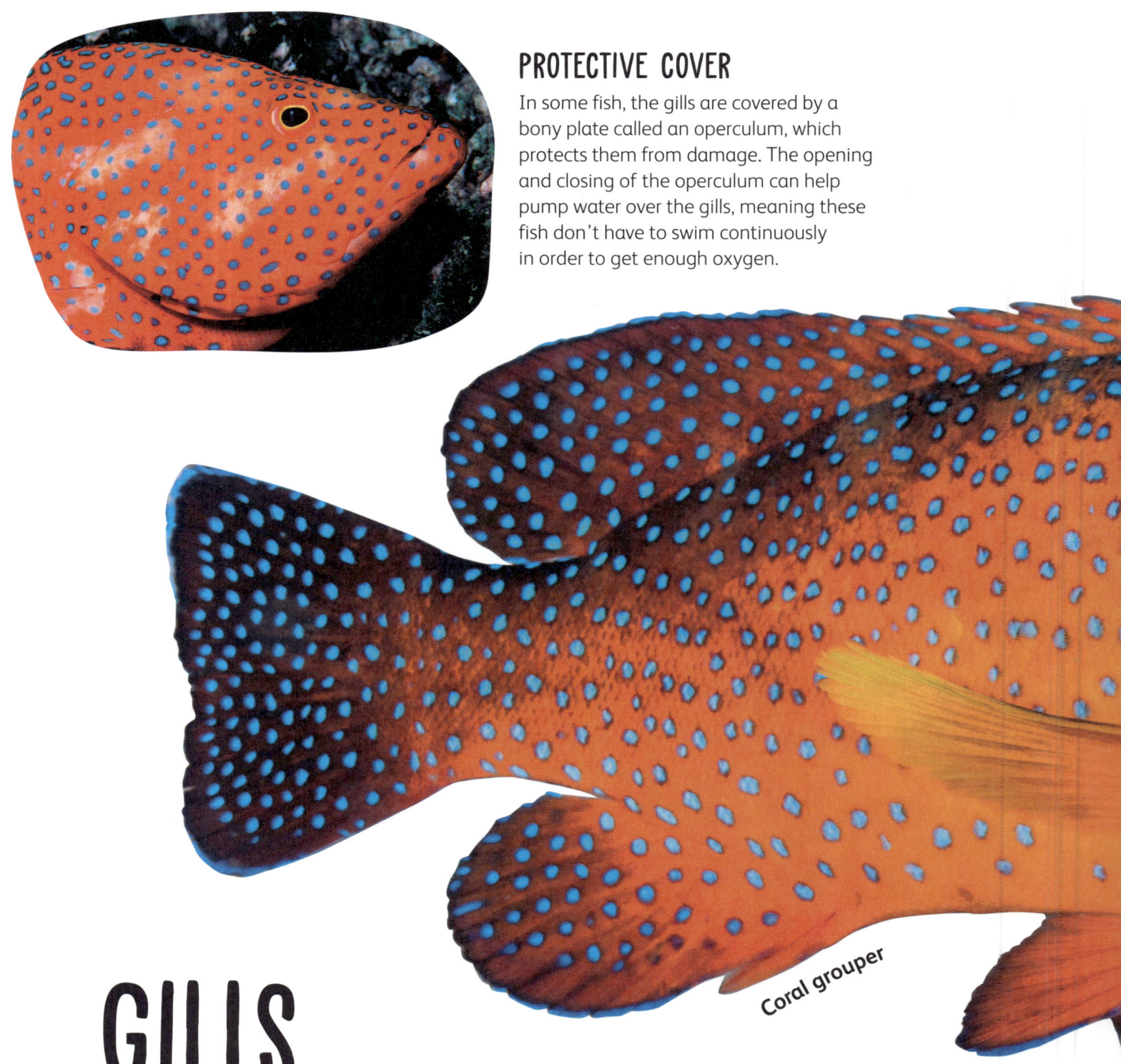

Coral grouper

GILLS

You may find it surprising, but **oxygen** gas can dissolve in water. However, water holds much less oxygen than air. Fish have **evolved** brilliant tools for extracting the oxygen they need to breathe – **organs** called **gills**. Usually, water moves over the gills in one direction only. First, water is drawn into the fish's mouth, next it moves over the gills where the **blood** is **oxygenated**, and then it exits behind the head.

FISH ON LAND

A handful of fish have evolved the ability to extract oxygen from air as well as water. Some have **lungs**, but not mudskippers. They can absorb oxygen through their skin or hold water droplets in their gills, meaning they can stay out of the water long enough to hop across the land – and even climb trees!

There are five gills on either side of the head, each made up of a sheet of filaments.

Each gill sheet is supported by a curved bone called a gill arch.

Gill rakers are bony spikes on the inside of the gill arches. They prevent objects damaging the gills and help catch food.

When a fish opens its mouth, water is drawn in.

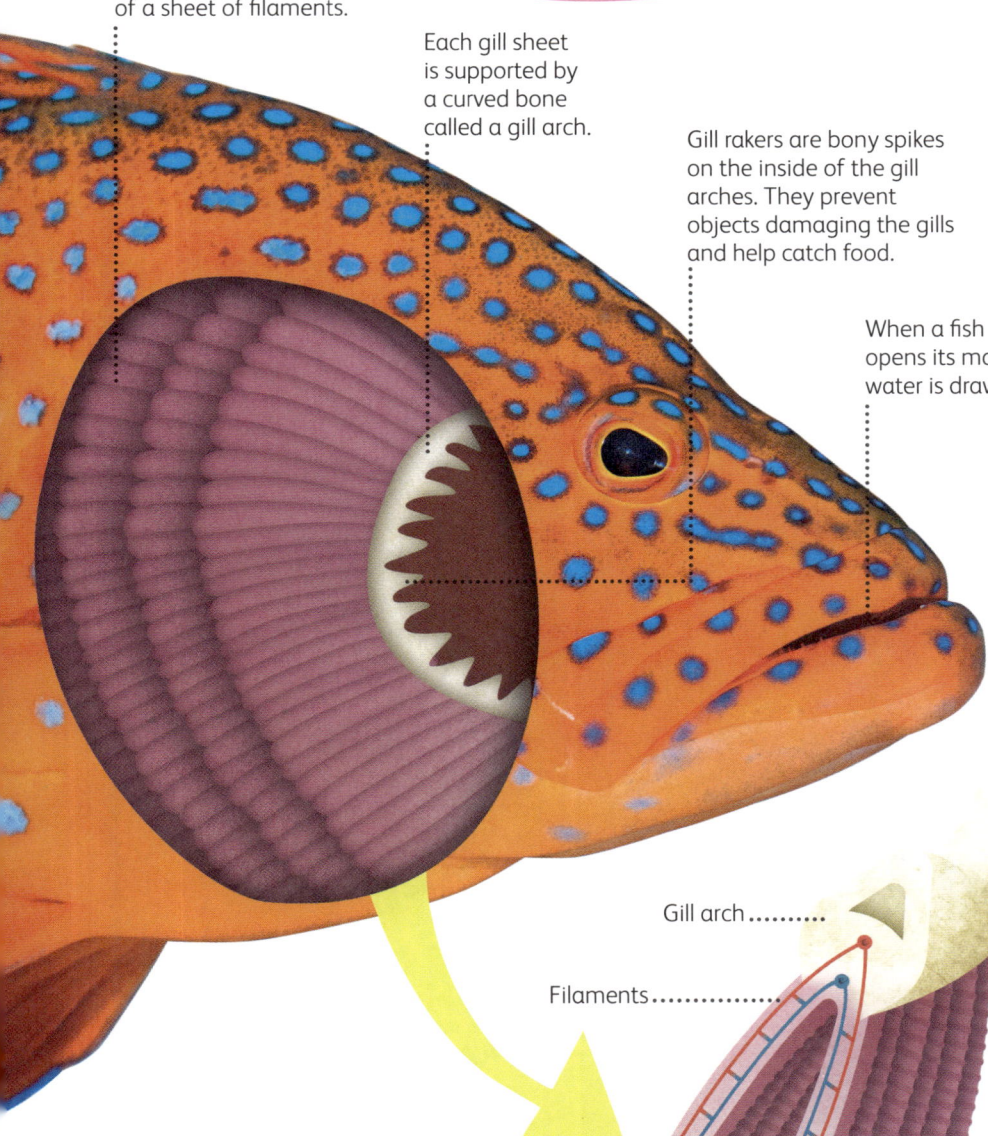

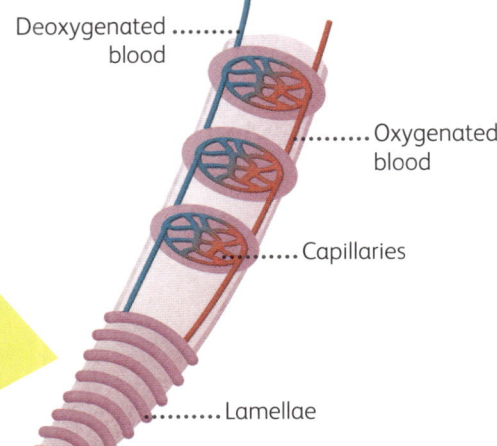

GILL CLEANING

Since they do such an important job, it's essential for fish to keep their gills clean. Unfortunately the ocean is full of **parasites** which love to feed on the blood-rich gills of fish. Some fish team up with cleaner shrimp, which remove the parasites and gobble them up – so everyone wins!

Gill arch

Filaments

Deoxygenated blood

Oxygenated blood

Capillaries

Lamellae

GILL STRUCTURE

Each gill sheet is made up of paired rows of structures called filaments. When water flows over the filaments, oxygen moves into **capillaries** just below their surface. The filaments contain many folds called lamellae, which are just one **cell** thick, making it easier for oxygen to pass into the blood.

BIRD LUNGS

Birds have a very efficient way of getting **oxygen** out of the air, which takes two breaths in and out to fully complete. Their system only allows air to move in one direction through the body and requires air to be stored in **air sacs** connected to the **lungs**. This prevents "old" and "fresh" air from mixing, making it easier for the lungs to extract as much oxygen as possible.

Arctic tern

Most birds have nine air sacs attached to the lungs. They take up most of the space inside a bird's body.

The lungs are firmly attached to the body wall and do not inflate and deflate like those of **mammals**.

Birds have no **diaphragm**. Their breaths are powered by **muscles** between the ribs.

BIRD BREATHING

When a bird first inhales, oxygen-rich air travels past the lungs to the middle and rear air sacs. When it exhales, air moves from those air sacs into the lungs, where oxygen is absorbed. When the bird inhales for a second time, the old air moves from the lungs to the forward air sacs and on the second exhale air leaves through the **trachea**.

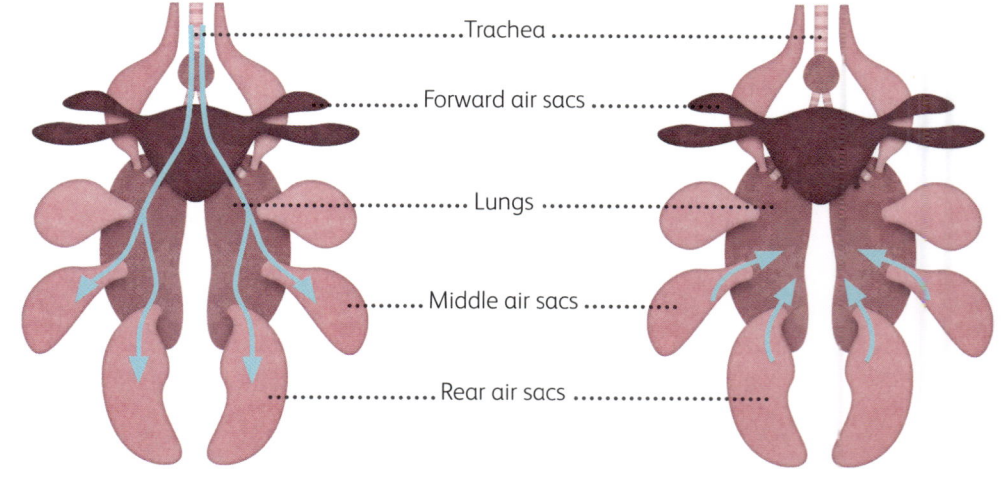

First inhale · First exhale

Trachea · Forward air sacs · Lungs · Middle air sacs · Rear air sacs

MIGRATION

Air sacs act like balloons, making a bird's body light. This an important **adaptation** when it comes to flight, especially over long distances, such as when the Arctic tern **migrates** from the Antarctic to the Arctic to breed. The ability to extract oxygen from the air efficiently also allows birds to breathe at higher altitudes, where the air contains less oxygen.

AIR CUSHIONS

Gannets, and some other seabirds, have extra air sacs under their skin. Gannets can dive into water at speeds of up to 97 kph (60 mph) in search of fish! It's thought that these air sacs may act as cushions, protecting the birds' bodies from the impact of hitting the water.

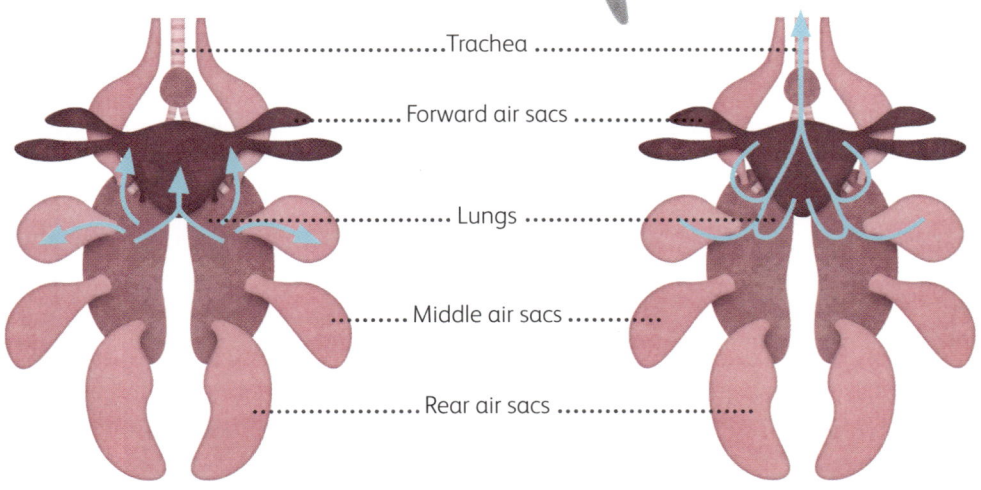

- Trachea
- Forward air sacs
- Lungs
- Middle air sacs
- Rear air sacs

Second inhale

Second exhale

When birds inhale, air leaves the lungs and when they exhale, air enters the lungs.

67

RESPIRATION

WHALE BLOWHOLE

It's hard to believe that the holes on top of a whale's head are its nostrils, but it's true! In most animals the trachea leads from the lungs to the mouth, but in whales and dolphins, the trachea leads to the top of the head. This means these animals can breathe without having to lift their face out of the water.

MAMMAL LUNGS

Inside their chest, protected by the ribcage, **mammals** have two **organs** called **lungs**, which are used for breathing. Lungs are filled with tiny **air sacs** that are brilliant at extracting **oxygen** from the air, however, this means that any mammal which spends time underwater must return to the surface to take a breath. So how do sea lions keep their bodies supplied with oxygen when diving beneath the waves?

HUNTING UNDERWATER

Sea lions are brilliant hunters, diving hundreds of metres under the surface in pursuit of their **prey**. To travel such distances, they must be able to hold their breath for many minutes at a time. Their heart rate also slows, reducing the amount of oxygen that is being pumped around the body, and they divert blood from the skin to their important organs, such as the **brain** and **heart**.

California sea lion

The **diaphragm** is a dome-shaped **muscle**. When it contracts the chest expands, pulling air into the lungs.

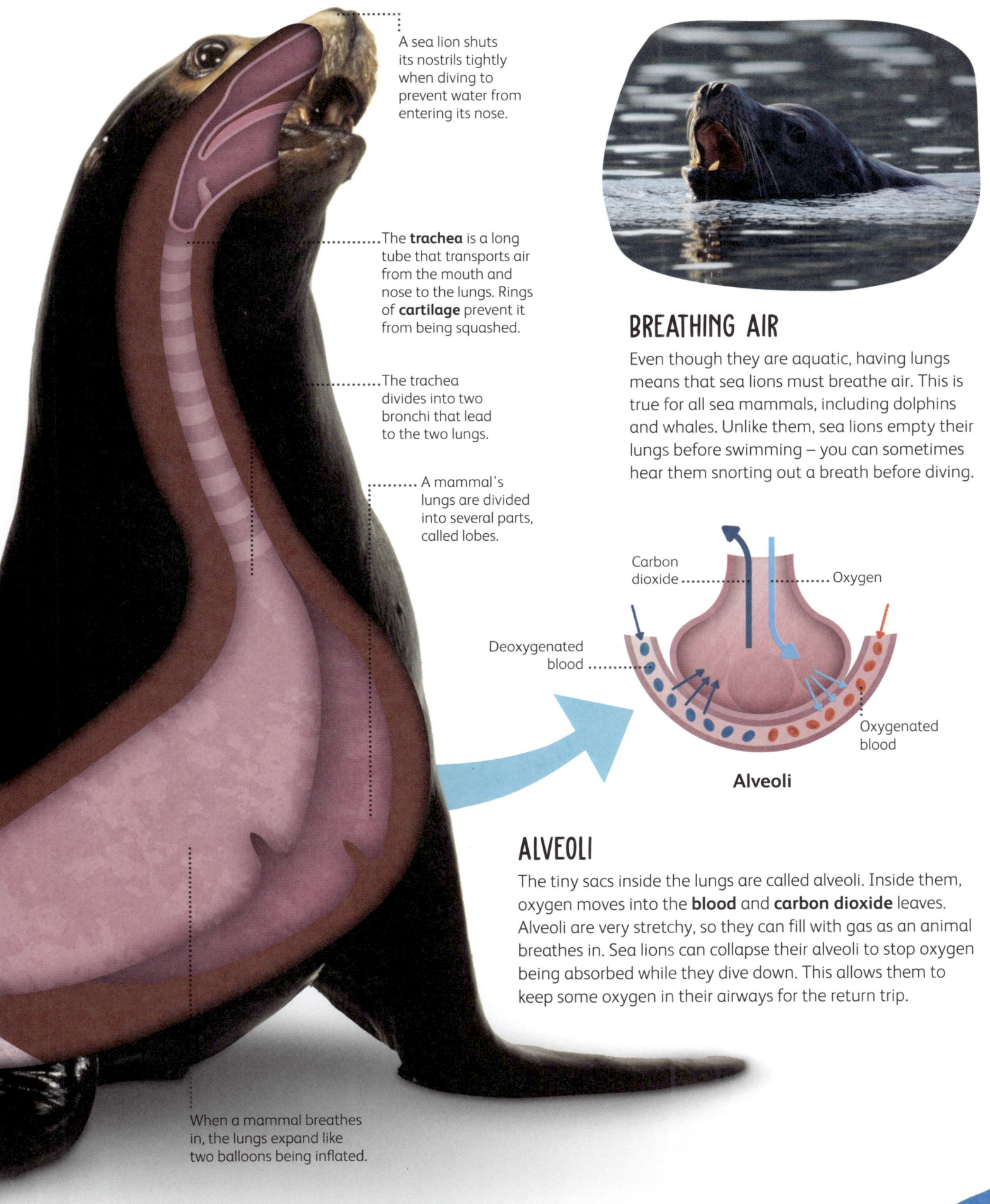

A sea lion shuts its nostrils tightly when diving to prevent water from entering its nose.

The **trachea** is a long tube that transports air from the mouth and nose to the lungs. Rings of **cartilage** prevent it from being squashed.

The trachea divides into two bronchi that lead to the two lungs.

A mammal's lungs are divided into several parts, called lobes.

When a mammal breathes in, the lungs expand like two balloons being inflated.

BREATHING AIR

Even though they are aquatic, having lungs means that sea lions must breathe air. This is true for all sea mammals, including dolphins and whales. Unlike them, sea lions empty their lungs before swimming – you can sometimes hear them snorting out a breath before diving.

Alveoli

Carbon dioxide

Oxygen

Deoxygenated blood

Oxygenated blood

ALVEOLI

The tiny sacs inside the lungs are called alveoli. Inside them, oxygen moves into the **blood** and **carbon dioxide** leaves. Alveoli are very stretchy, so they can fill with gas as an animal breathes in. Sea lions can collapse their alveoli to stop oxygen being absorbed while they dive down. This allows them to keep some oxygen in their airways for the return trip.

SOUND

Animals make sounds for all kinds of reasons, such as to express emotions, to communicate with other animals, to attract **mates**, or to raise the alarm. Some animals even use sound to catch food and find their way. There are lots of different ways to make sounds – and many of these use the same systems that are used for breathing.

LIFE SIZE

COMMON COQUÍ FROG
The coquí frog from Puerto Rico is one of the smallest tree frogs in the world. However, a sac that inflates like a balloon at the base of its mouth amplifies its call to make it as loud as a motorbike!

Vocal cords
Amphibians, **reptiles**, and **mammals** make sounds using an **organ** at the top of the **trachea** called the larynx, which contains folds called vocal cords. When air from the **lungs** is forced over the vocal cords, they vibrate, making a noise.

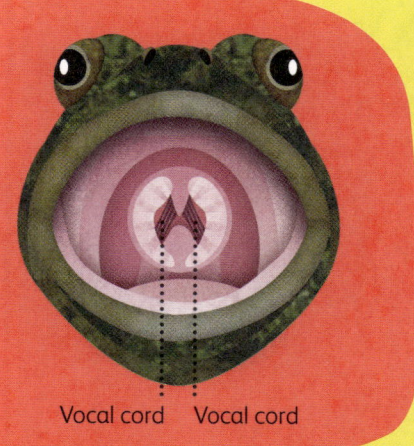

Vocal cord   Vocal cord

COBRA
The cobra has chambers of air along its trachea, which allow it to make a growing sound that is much lower in pitch than the hissing of other snakes. Hissing is produced by pushing air out of an opening in the mouth called the glottis.

PARROT

Instead of a larynx, birds have a sound-making organ called a syrinx. A parrot's syrinx is incredibly complex, which allows it to mimic a wide range of sounds, including human speech!

DOLPHIN

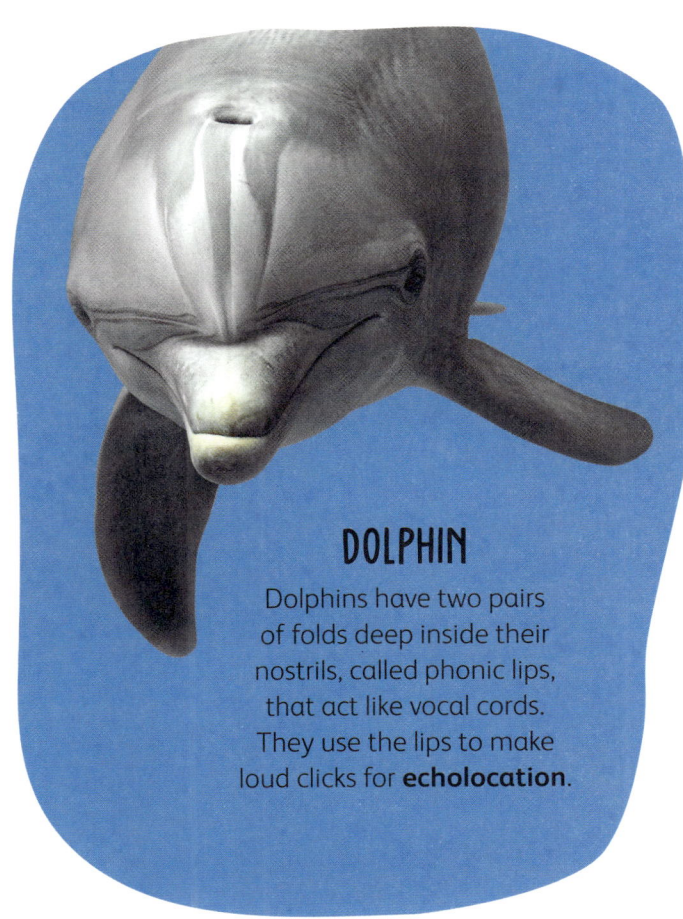

Dolphins have two pairs of folds deep inside their nostrils, called phonic lips, that act like vocal cords. They use the lips to make loud clicks for **echolocation**.

COYOTE

Members of the dog family, such as coyotes, can make a wide range of noises including yelps, whimpers, whines, moans, growls, snarls, barks, and howls. They mostly use sound to communicate with their group, or pack.

TIGER

Tigers, lions, leopards, and jaguars have a special stretchy **ligament** in place of one of the **bones** of the larynx. This allows them to widen the larynx and make a deep-pitched growl called a roar.

71

RESPIRATION

INSIDE A SNAKE

Pythons are big, powerful snakes, which kill **prey** by wrapping it up in their strong, muscular bodies and squeezing it tightly. Snakes cannot chew their prey, which is often very large, so they swallow it whole! All that crushing and eating huge meals could make it hard for pythons to breathe, were it not for some special **adaptations**.

BREATHING TUBE

A hole in a snake's mouth, called the glottis, leads to the **trachea**. When the snake is eating, it can push the glottis out of the side of its mouth to breathe through, like a snorkel. The glottis is also what makes the hissing sound when a snake breathes out heavily.

The fleshy front section of the right lung is where **oxygen** is absorbed.

The left lung is very small in pythons. It is not used for absorbing oxygen. Organs like this that no longer function are known as vestigial.

LARGE LUNG

Snakes have only one large **lung**, the second lung is either missing or very tiny. The front section of the large lung is where **gas exchange** happens. The back section, however, is an **air sac**. This part stores air and helps draw it through the front part when it can't inflate fully.

RIB MUSCLES

Snakes do not have a **diaphragm**. Instead, their breathing is controlled by **muscles** between the ribs, which expand and contract the ribcage, and the large lung inside it. In snakes that squeeze their prey, this is the work of the intercostal muscles between the ribs, and the levator costae muscles, which attach the ribs to the **backbone**.

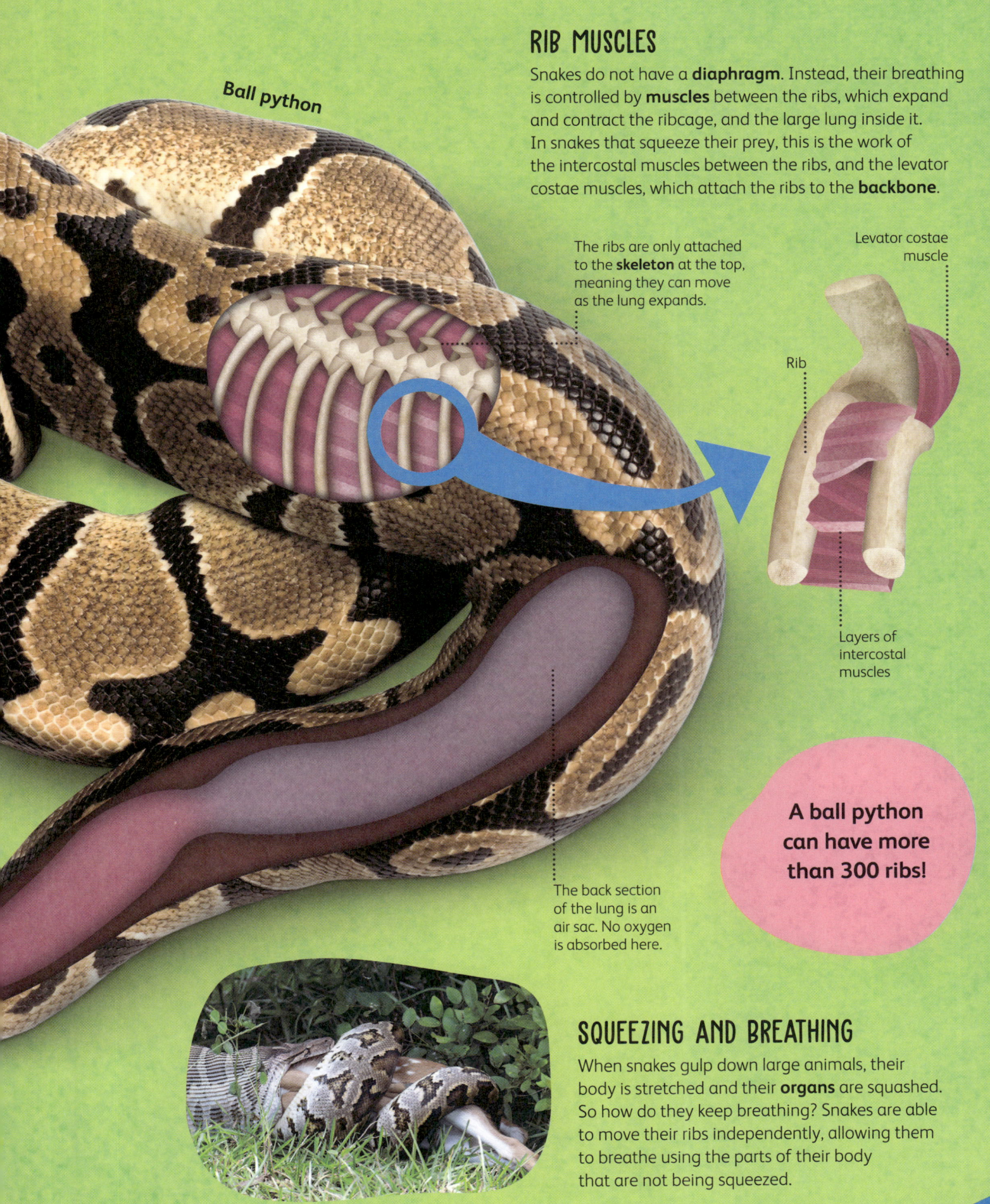

Ball python

The ribs are only attached to the **skeleton** at the top, meaning they can move as the lung expands.

Levator costae muscle

Rib

Layers of intercostal muscles

The back section of the lung is an air sac. No oxygen is absorbed here.

A ball python can have more than 300 ribs!

SQUEEZING AND BREATHING

When snakes gulp down large animals, their body is stretched and their **organs** are squashed. So how do they keep breathing? Snakes are able to move their ribs independently, allowing them to breathe using the parts of their body that are not being squeezed.

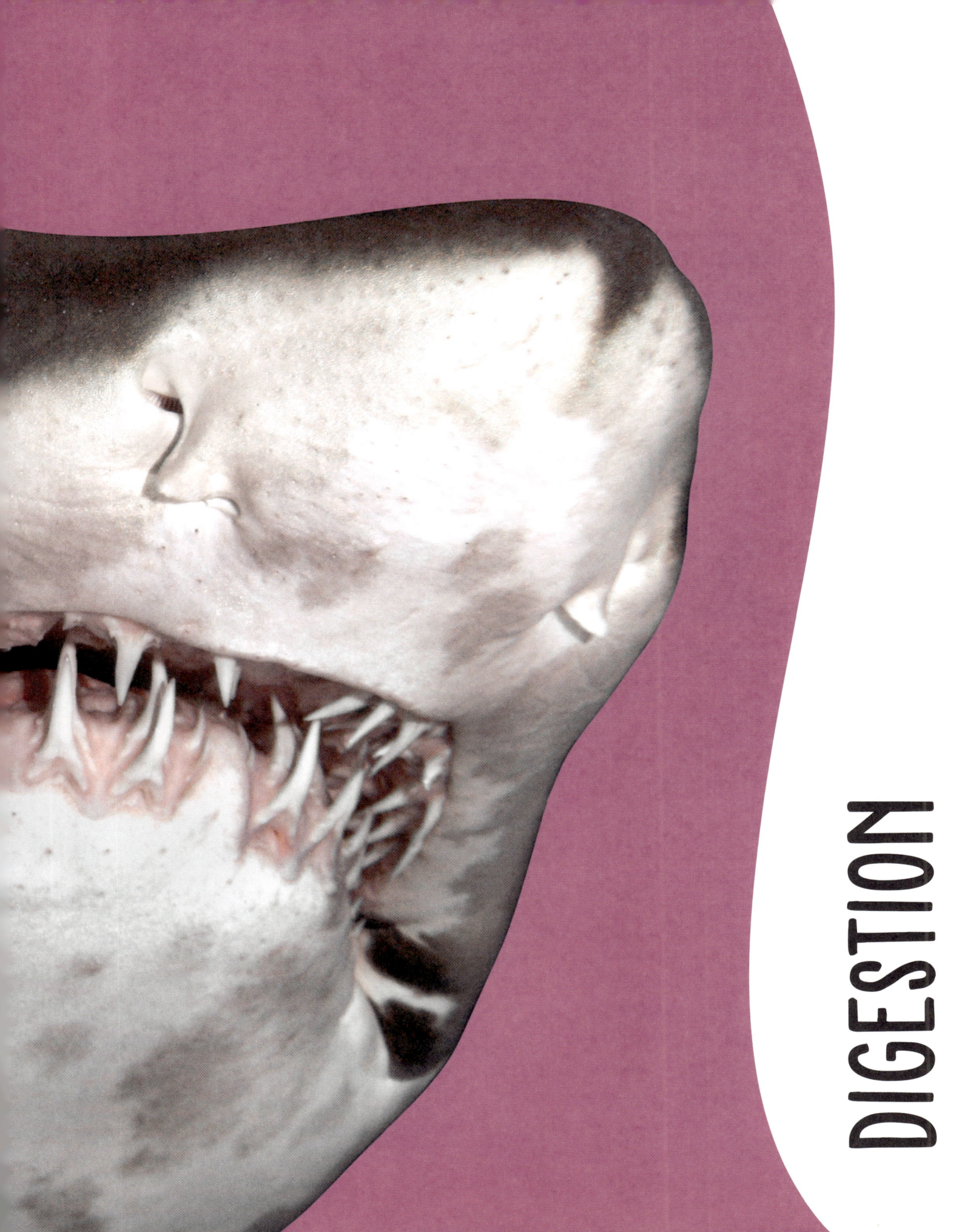

DIGESTION

WHAT IS DIGESTION?

In its simplest form, the digestive system is just one long tube from the mouth to the anus. As food moves through this tube, it is broken down and the parts can be absorbed by the body. The rest is released, or **excreted**, as waste. Breaking down food can be quite a difficult task, so animals have a whole group of **organs** dedicated to this job.

MAMMAL'S DIGESTIVE SYSTEM

The basic layout of the digestive system is similar in all **mammals**, but the organs within it often have specific **adaptations** depending on the animal's diet. Raccoons are **omnivores**, which means they eat both plants and meat.

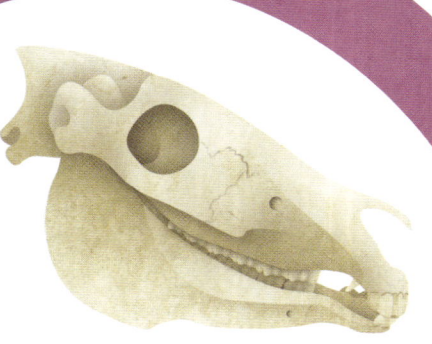

Herbivore skull

Herbivores, such as zebras, eat only plants. Their back teeth, or molars, are often wide and flat. This helps them grind down the tough plant matter they eat very finely before it is swallowed.

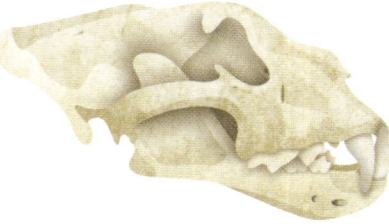

Carnivore skull

Carnivores, such as jaguars, eat only meat. They often have long, pointed canine teeth, which are used for gripping **prey** and tearing flesh from **bones**. They can swallow large chunks of food without chewing.

Teeth
Raccoons have sharp front teeth to slice through meat and flattened back teeth for crunching nuts and grains.

Mouth
In the mouth, watery **saliva** is mixed with food to help it move easily through the digestive system.

Tongue
Many animals use their tongues to taste foods. This can help them to decide which foods are good to eat.

Oesophagus
The **oesophagus** is a muscular tube connecting the back of the mouth to the stomach.

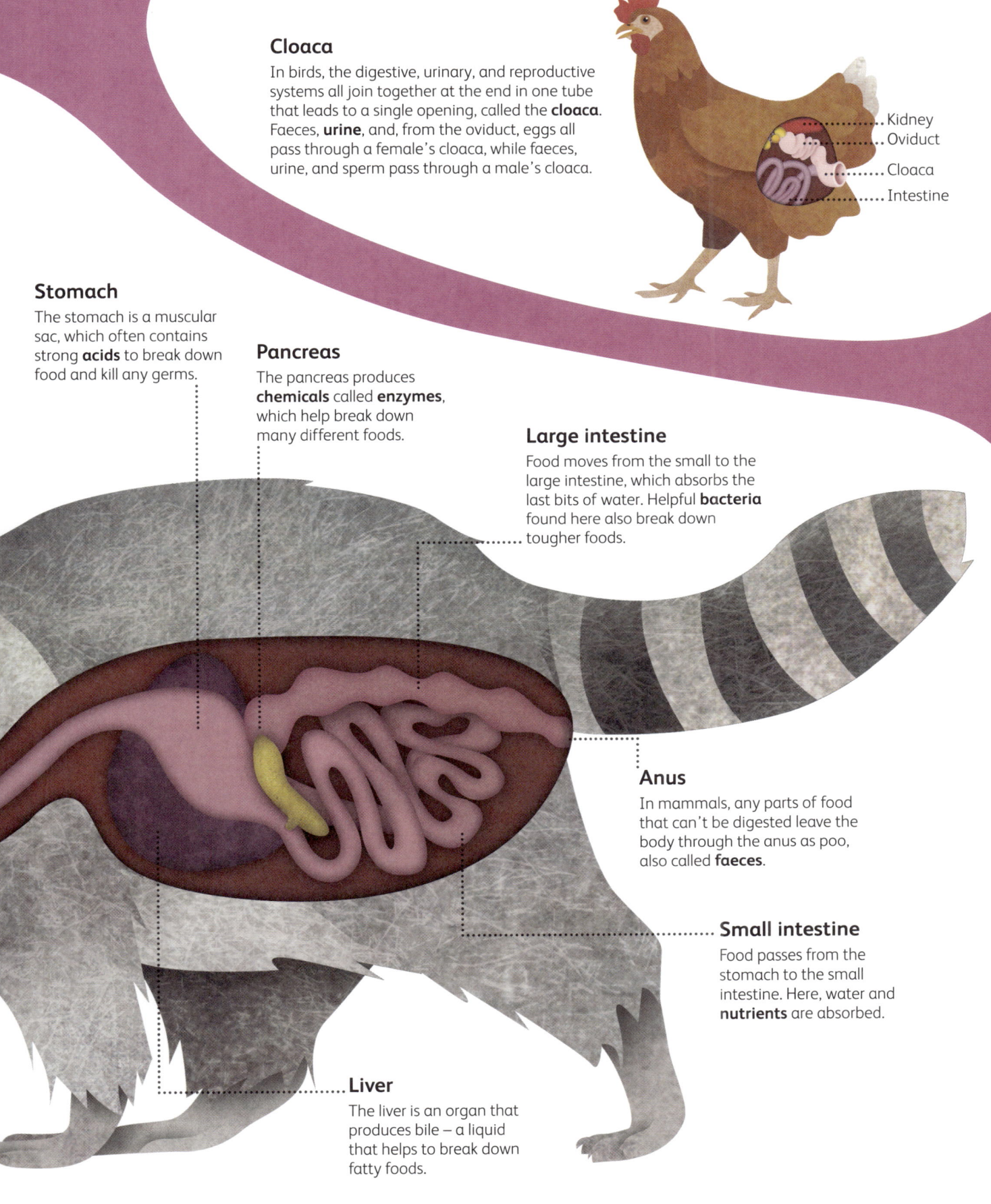

Cloaca

In birds, the digestive, urinary, and reproductive systems all join together at the end in one tube that leads to a single opening, called the **cloaca**. Faeces, **urine**, and, from the oviduct, eggs all pass through a female's cloaca, while faeces, urine, and sperm pass through a male's cloaca.

Kidney
Oviduct
Cloaca
Intestine

Stomach

The stomach is a muscular sac, which often contains strong **acids** to break down food and kill any germs.

Pancreas

The pancreas produces **chemicals** called **enzymes**, which help break down many different foods.

Large intestine

Food moves from the small to the large intestine, which absorbs the last bits of water. Helpful **bacteria** found here also break down tougher foods.

Anus

In mammals, any parts of food that can't be digested leave the body through the anus as poo, also called **faeces**.

Small intestine

Food passes from the stomach to the small intestine. Here, water and **nutrients** are absorbed.

Liver

The liver is an organ that produces bile – a liquid that helps to break down fatty foods.

DIGESTION

MOUTH

The mouth is where **digestion** begins. Here, food enters the body and is ripped, chewed, or snipped into smaller pieces to make it easier to swallow. Some animals have teeth for this job, but others – like turtles – have a beak. The tongue can also play an important role in moving food around the mouth, while **saliva** makes it wet and easier to swallow.

The hawksbill turtle has a sharp, pointed beak, which it uses to pull sponges from small cracks in rocks.

SLICING BEAK

In place of teeth, turtles have extremely strong beaks, which are made from tough **keratin** that grows over plates of **bone**. Adult green sea turtles are **herbivores** and have beaks with serrated edges, which are perfect for tearing up seagrass and rasping algae off hard surfaces.

FILTER FEEDING

Some types of whale don't have teeth and can't chew. These are baleen whales. Instead, they get their food by taking in huge gulps of sea water and filtering out small sea creatures and plankton using a dense mat of strong, bristle-like hairs, called baleen, which are attached to their upper **jaws**.

JELLY SNACK

Jellyfish are slippery creatures that often come with a **venomous** sting, but that doesn't stop some sea turtles from gobbling them up! The throats of these sea turtles are lined with backwards-pointing spikes, which prevent jellyfish from slipping back up the turtle's throat. This thick lining also prevents the turtle from being stung.

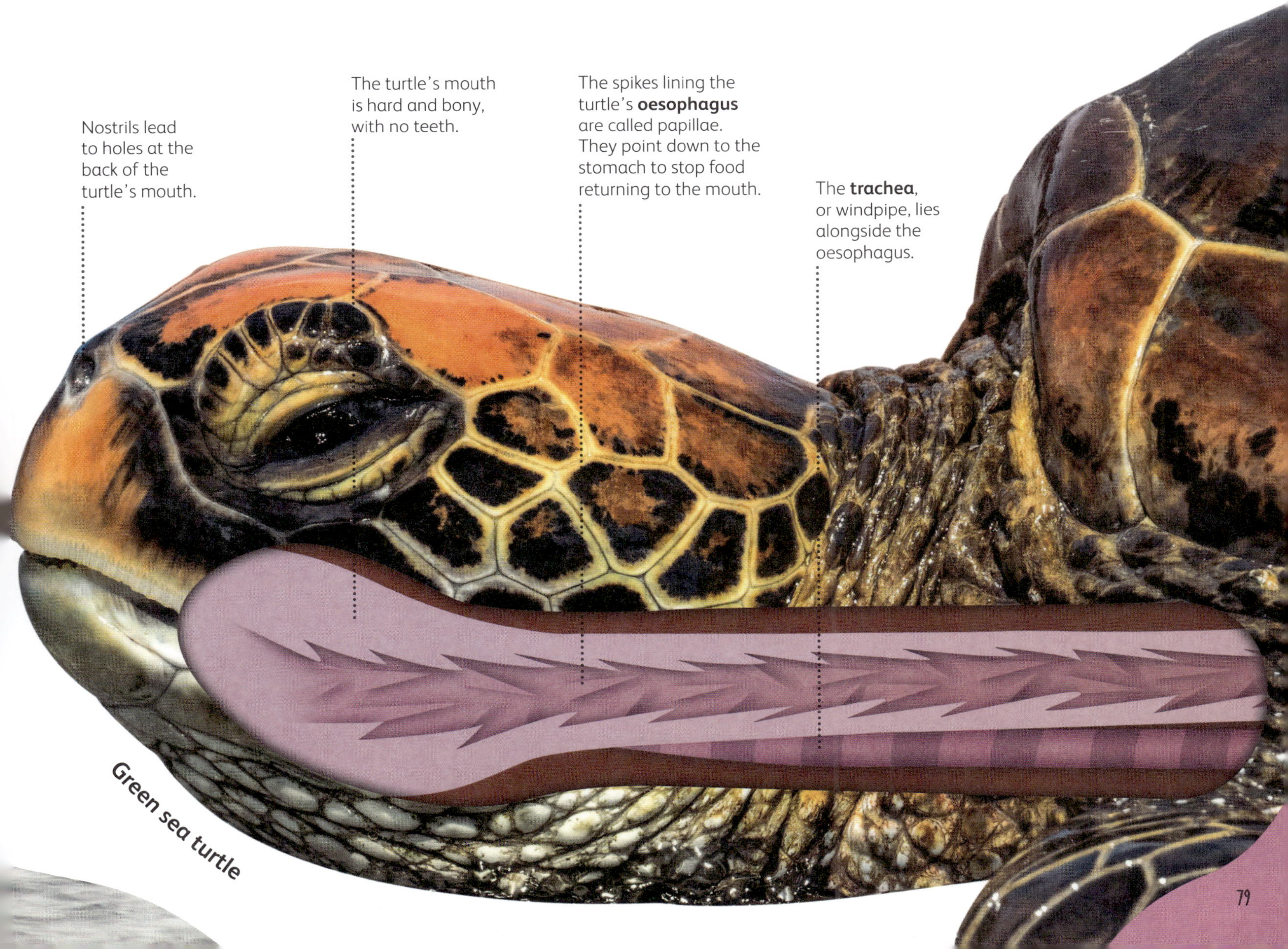

Nostrils lead to holes at the back of the turtle's mouth.

The turtle's mouth is hard and bony, with no teeth.

The spikes lining the turtle's **oesophagus** are called papillae. They point down to the stomach to stop food returning to the mouth.

The **trachea**, or windpipe, lies alongside the oesophagus.

Green sea turtle

DIGESTION

STOMACH

When food is swallowed, it travels down the **oesophagus** to the stomach. The stomach is a muscular sac that stores food before it moves down into the **intestines**, so its walls must be very stretchy to accommodate large meals. Bison belong to a group of animals known as **ruminants**, which feed on plants and have a specialized stomach that is split into four chambers to help break down tough leaves.

The first and largest stomach chamber is the rumen. It is filled with **microbes** that break down food and release stinky gases.

Muscles in the oesophagus help to move grass from the stomach back to the mouth to be chewed again.

The second chamber is the reticulum. It is easily identified by its walls, which are patterned like honeycomb.

Bison have large, flattened molar teeth for grinding down plant material.

American bison

GRAZING ON GRASS

Bison are grazers, meaning most of their diet is made up of grasses. While grass is pretty easy to find, it is not very high in energy, so bison must eat a lot of it to get all of the **nutrients** they need to survive.

CHEWING THE CUD

Grass is difficult to digest. **Herbivores** have evolved a range of different ways to deal with this problem – ruminants chew it twice! After the first round of chewing, grass moves into the rumen, where it mixes with **bacteria** and other microbes before being regurgitated into the mouth as cud, ready to be chewed again.

The soup of broken-down grass moves from the abomasum to the intestines, where water and nutrients are extracted.

The fourth chamber is the abomasum. This is more like a typical **mammal's** stomach and contains **acids** and **enzymes** to break down food.

The third chamber is the omasum, its walls are folded like the pages in a book. Here, water is absorbed.

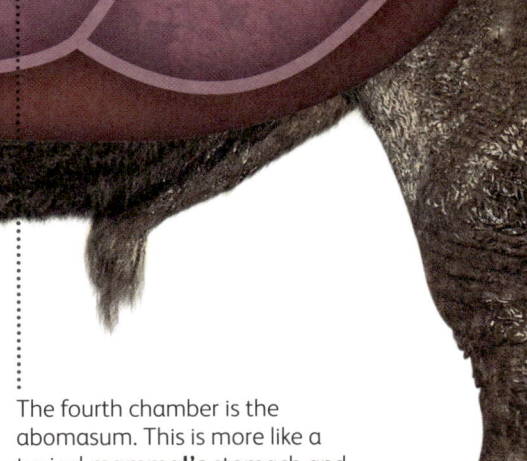

NO STOMACH

Seahorses have no teeth or stomach! Their digestive system is simply one long tube, and it is not very good at breaking down food. As food passes through their body so quickly, seahorses are almost always eating, sucking in small crustaceans through their tube-shaped mouths.

DIGESTION

LIVER

The liver is huge! In many animals, it is the biggest **organ** after the skin. It is always very busy with lots of important jobs, including filtering out and breaking down harmful **toxins**, storing **vitamins** and **minerals**, making useful **chemicals**, and fighting infections. It also helps the digestive system to work. In animals that sleep during the winter, the liver must also change how **nutrients** in the body are processed.

Brown bear

LIGHT LIVER

Sharks have enormous, oily livers which can make up a quarter of their entire weight and take up the majority of their body. As well as providing a source of energy, the oils within a shark's liver make it less dense than water, helping the shark to stay afloat.

WINTER BREAK

During the winter months when there is less food available, some species of bear, including brown bears, go into a deep sleep known as hibernation. For up to eight months they don't eat, their heartbeat slows, and their temperature drops. While the bear is hibernating, its extra-strong bile helps to break down **fats** and toxins.

CHEMICAL FACTORIES

The digestive system uses lots of special chemicals to break down food. Many of these are produced by the liver and pancreas. The liver produces bile, a green liquid that breaks down fats. It is stored in a small, saclike organ called the gallbladder. The pancreas produces **enzymes** – chemicals that break down **sugars**, fats, and **proteins**.

Liver

Pancreas

Gallbladder   Small intestine

The liver sits directly below the **diaphragm**, a dish-shaped **muscle** that helps with breathing.

DIGESTION

DROPPINGS FOR DINNER

The waste products of digestion leave the body as poo, or **faeces**. For most animals, that is the end of the process, but not for rabbits! In order to get every last bit of nutrition from their food, they produce a special type of poo called a caecotroph, which they eat and digest again!

INTESTINES

Once food has passed through the stomach, it moves into a very long, muscular tube known as the **intestine**, where **digestion** continues. Inside the intestine, **nutrients** and water are absorbed, and waste products are formed into poo. The intestine is split into two main sections: the small intestine and the large intestine. The bulk of a rabbit's digestion takes place in the large intestine, in a special organ called the caecum.

ALL-DAY DIGESTION

Koalas feed on the tough leaves of eucalyptus trees, which are poisonous to most animals. Just like rabbits, most of their digestion occurs in the caecum. The food they eat is so fibrous and difficult to digest, they spend up to 22 hours of the day sleeping!

It takes about 19 hours for food to pass through a rabbit's digestive system.

TOUGH PLANTS

Plants are full of **fibre**, which is a tough material that is hard to break down. This means plants are much more difficult to digest than meat. **Herbivores** have a range of special **adaptations** to help with this. Rabbits have teeth that never stop growing, so they do not get worn down through long days of munching grass.

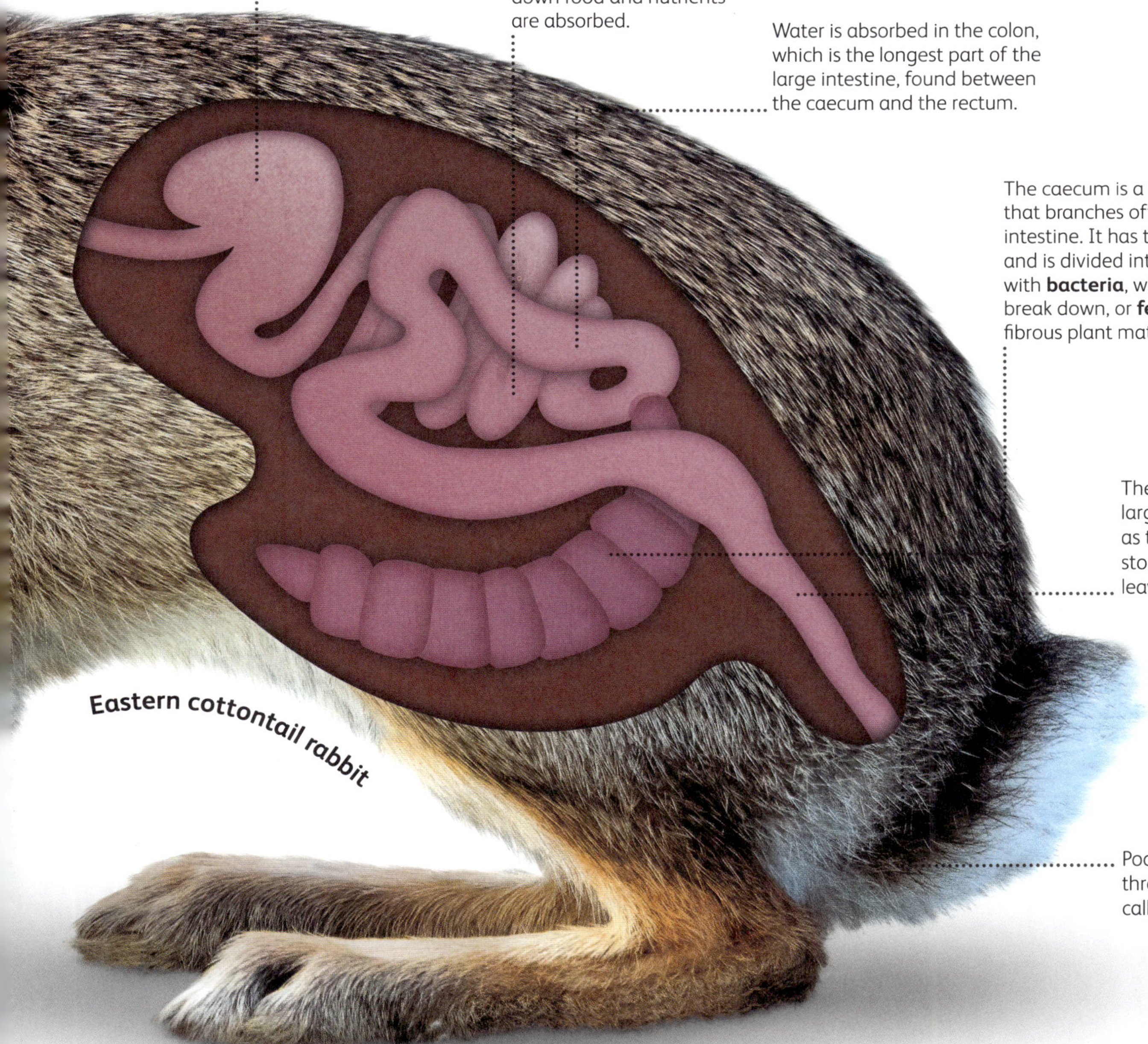

Rabbits have a simple stomach with just one chamber. It is filled with **acid** to kill germs.

In the small intestine, **enzymes** help to break down food and nutrients are absorbed.

Water is absorbed in the colon, which is the longest part of the large intestine, found between the caecum and the rectum.

The caecum is a pouch that branches off the large intestine. It has thin walls and is divided into sacs filled with **bacteria**, which help to break down, or **ferment**, fibrous plant material.

The final stretch of the large intestine is known as the rectum. Poo is stored here until it leaves the body.

Poo exits the body through an opening called the anus.

Eastern cottontail rabbit

DIGESTION

SNEEZING SALT

Salt **glands** are small organs found in the noses of marine iguanas and seabirds that allow them to sneeze out extra salt through their nostrils. This means less work for the kidneys, which is particularly important in **reptiles** as their kidneys cannot produce urine that contains more waste than their blood.

The salty fluid from the salt glands drips down into the nostrils to be removed.

KIDNEYS

As **blood** travels around the body it collects waste. Removing this waste is the job of the kidneys, a pair of **organs** which filter the blood, keeping hold of the useful bits and mixing unwanted materials with excess water to form **urine** or similar waste products. When animals such as marine iguanas feed in seawater, the waste filtration system faces an extra challenge – getting rid of all that salt.

DRINKING PATCH

Most animals must drink water to stay hydrated, but frogs have an area of skin on their belly called a drinking patch, which absorbs water directly into the blood. Water is important for creating fluids in the body and helping to remove waste from the blood.

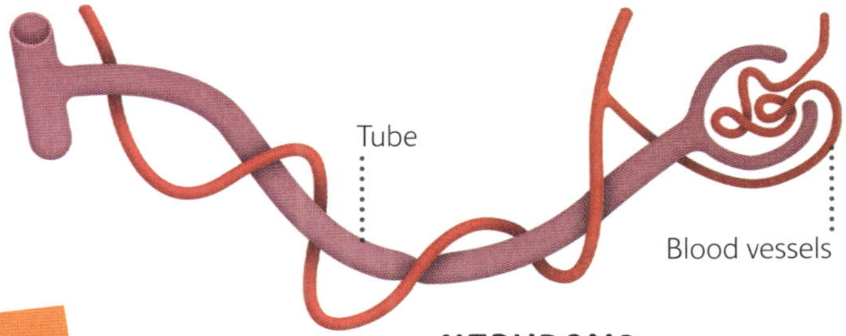

Marine iguana

Tube

Blood vessels

NEPHRONS

There are thousands of tiny filters within each kidney, called nephrons. Blood is pushed into the nephrons at high pressure, causing salts, waste, and **nutrients** to be pushed into a special tube. The important bits are reabsorbed, leaving just the waste products to be passed to the bladder.

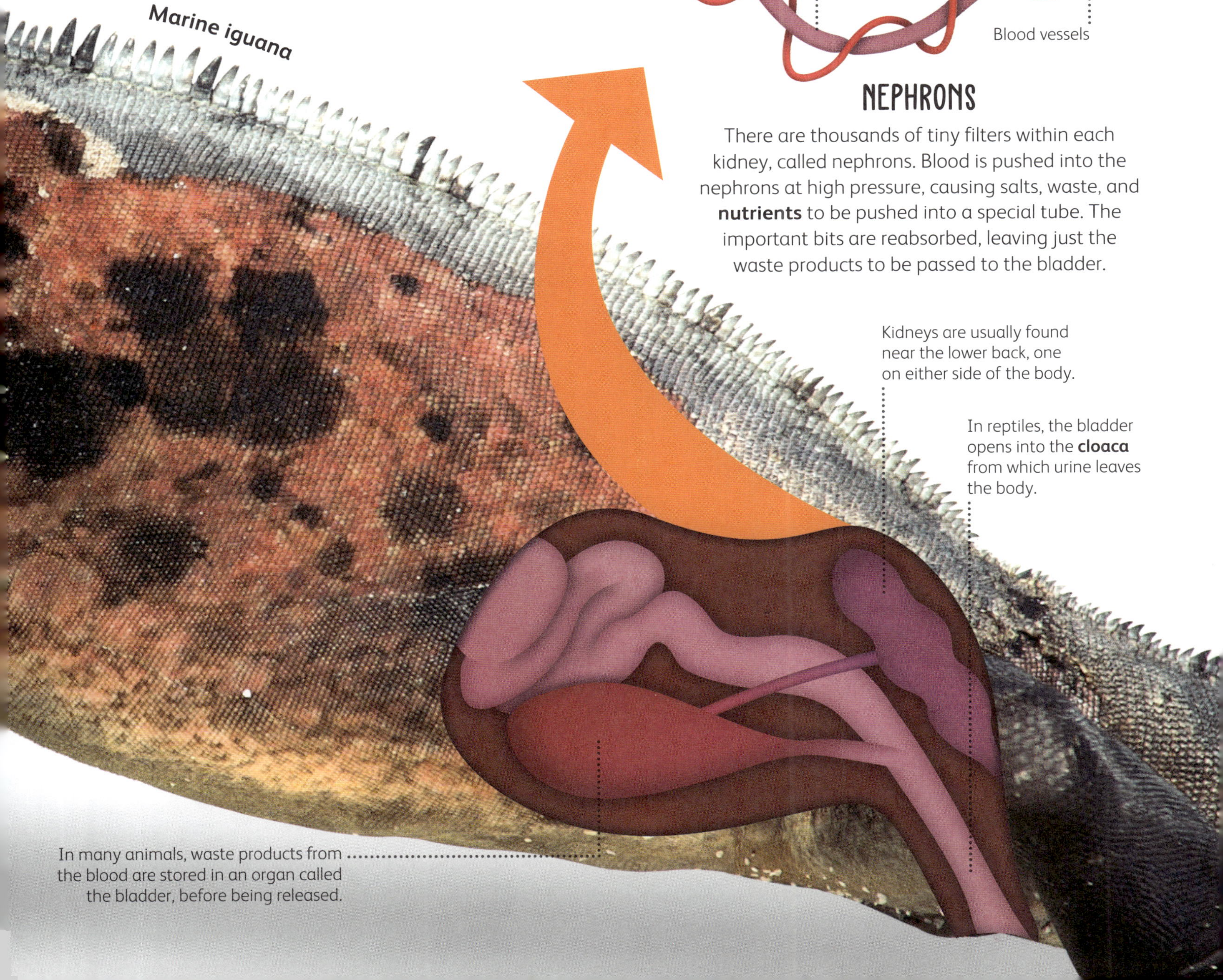

Kidneys are usually found near the lower back, one on either side of the body.

In reptiles, the bladder opens into the **cloaca** from which urine leaves the body.

In many animals, waste products from the blood are stored in an organ called the bladder, before being released.

DIGESTION

TEETH

Teeth come in many shapes and sizes. They can be used for chewing, biting, tearing, crushing, and grinding, and often play an important role in **digestion**, breaking down food into smaller pieces so it is ready to be swallowed. However, teeth can have some other fascinating jobs too.

SNAKE

Snakes have multiple rows of long, thin teeth like needles, which point backwards towards the **oesophagus**. These teeth are for grabbing, not chewing, meaning almost all snakes must swallow their prey whole. They also have two extra-long teeth that can deliver a deadly bite!

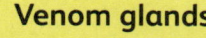
LIFE SIZE

Venom glands
Some snakes have fangs, which are two long, hollow teeth that act like needles to inject toxic **venom** into attackers or **prey**. The venom is made in **organs** called venom **glands** in the snake's cheeks.

SHARK

Shark teeth are constantly on the move. Older teeth are slowly pushed forwards, to the front of the mouth, and fall out. However, they are replaced by new ones, like a tooth conveyor belt.

LAMPREY

Lampreys are jawless fish. Instead of biting **jaws**, they have a round, sucker-like mouth full of many rows of sharp teeth, which they can use to suck the **blood** from other animals.

BEAVER

Beavers use their super-strong teeth to gnaw through tree trunks. Their teeth are orange because they contain **iron** – the same reason our blood is red!

ELEPHANT

Elephant tusks are just long teeth. They are found on both males and females and can be used for digging for food, marking territory, warning off rivals, and self-defence.

NARWHAL

It looks a bit like a horn, but the tusk of a narwhal – a type of whale – is actually a tooth. The narwhal uses it for stunning its prey. Though both males and females can grow tusks, the males' are much larger.

DIGESTION

TOUGH BEAK

Instead of teeth, birds have hard beaks made of **keratin**. Bird beaks come in a huge variety of shapes and sizes depending on their diet. Parrots have tough, curved beaks for cracking open nuts and seeds. Some also have bristly tongues for collecting nectar.

STORAGE AREA

Some birds, including chickens, parrots, and pigeons, have a stretchy pouch in their neck called a crop. This is an expanded section of the **oesophagus** which is used for storing food before **digestion**. Parrots can fill their crops quickly when food is plentiful and then fly off to safety, away from **predators**.

The **oesophagus** is a flexible tube that leads from the mouth to the crop.

When full, the stretchy crop makes the parrot's neck bulge out.

INSIDE A PARROT

There are more than 400 different **species** of parrot, but two things that most of them have in common is a strong beak and the ability to eat many different kinds of foods. With no hands for grabbing and a diet that contains tough nuts and seeds, parrots need a specialized digestive system and a smart way to collect their meals.

Many birds swallow small stones, called gastroliths, which help to break down their food.

REGURGITATING FOOD
When parrot chicks hatch, they cannot leave the nest to find food, so their parents must bring back their meals. The adults store the food for their young in their crop, before returning to the nest and regurgitating it directly into their chicks' mouths.

Food is broken down by **chemicals** in the proventriculus.

Food is broken down physically by the muscular walls of the gizzard.

TWO STOMACHS
Bird stomachs are split into two chambers – the gizzard and the proventriculus. First is the proventriculus, which is full of **glands** that release **acids** and **enzymes** to begin digestion. Next, food moves down into the thick-walled, muscular gizzard to be ground down into smaller pieces.

From the gizzard, food passes into the small intestine.

SENSES

WHAT ARE SENSES?

Animals use their senses to gather information about the world around them. There are five main senses, but some animals have many more than these. The information is gathered by special **organs**, which send messages about what they have found to the **brain** via nerves. The brain puts all this information together to build a picture of the animal's surroundings.

Hearing
Ears can sense movements called vibrations in air or water, known as sound waves. This is how animals hear.

FIVE SENSES
Most animals, including the fennec fox, have at least five senses, including hearing, smell, taste, sight, and touch, but they differ in how effective they are. Some animals are brilliant at smelling, others have incredible sight, while some specialize in feeling their way around the world.

Sight
Eyes detect light, which gives them information about the size, colour, and shape of objects around them.

Smell
Special skin cells inside the nose detect tiny particles of the stinky objects around them as smell.

Touch
The skin can sense pressure and movement as touch. Hairs, including whiskers, are attached to touch receptors too.

Taste
Like smelling, tasting happens when tiny particles of food come into contact with sensors on the tongue.

NERVES

Nerves are bundles of **cells** known as neurons, which send and receive information around the body. Neurons have long tails called axons, which pass information quickly from one part of the body to another. Some nerves carry information to the brain and others carry instructions from the brain.

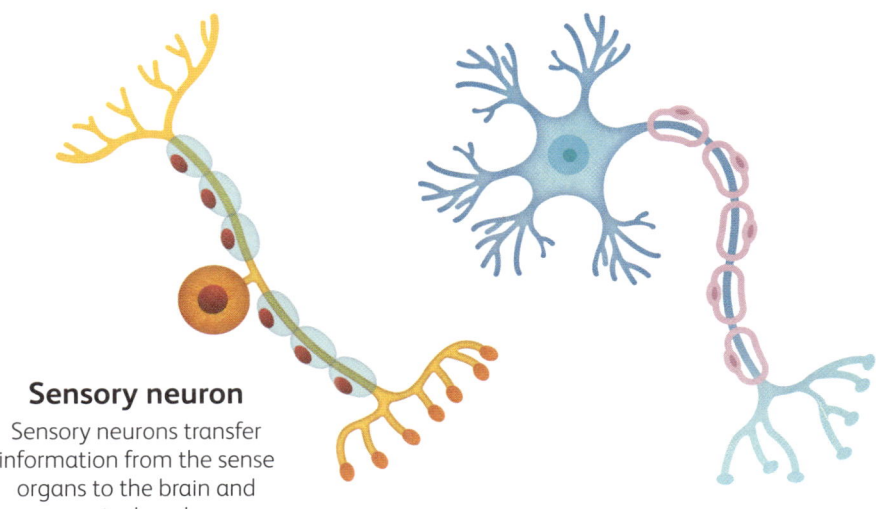

Sensory neuron
Sensory neurons transfer information from the sense organs to the brain and spinal cord.

Motor neuron
Motor neurons carry instructions from the brain and spinal cord to the **muscles** and other body parts to direct movements and other actions.

NERVOUS SYSTEM

The nervous system is made up of the brain, the **spinal cord**, and all the nerves that are attached to them. Nerves pass information to and from the brain in the form of electrical signals. Sometimes, signals only travel to the spinal cord, not the brain, before returning with a response – this is called a reflex.

Spinal cord
The spinal cord is a thick rope of nerves that runs the length of the body from the brain to the tail. It is protected by the **backbone**.

Brain
The brain is a mass of neurons and supporting cells. It processes information collected by nerves around the body.

Nerves
Nerves link up organs so that they can communicate with each other and the brain.

SENSES

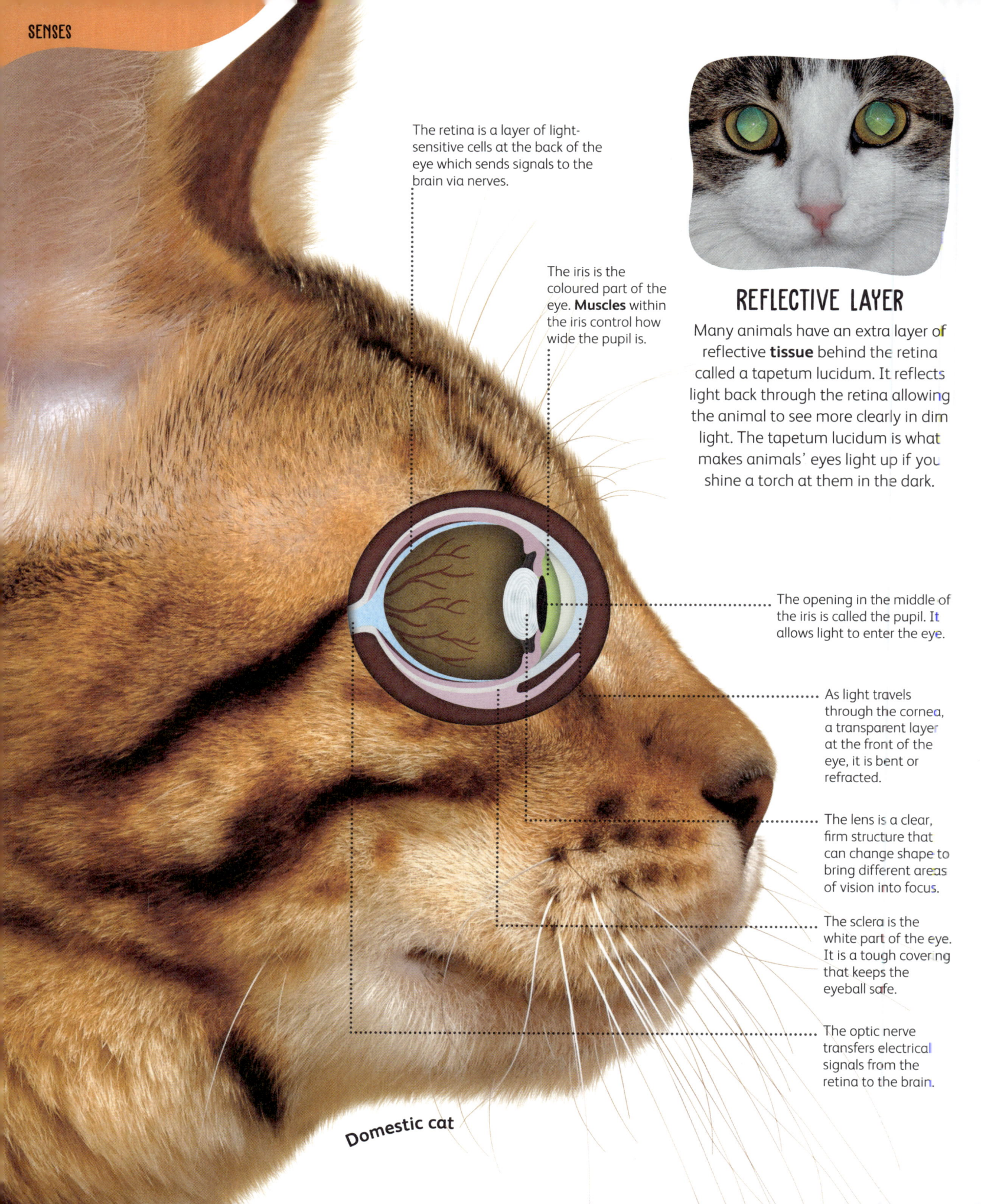

The retina is a layer of light-sensitive cells at the back of the eye which sends signals to the brain via nerves.

The iris is the coloured part of the eye. **Muscles** within the iris control how wide the pupil is.

REFLECTIVE LAYER

Many animals have an extra layer of reflective **tissue** behind the retina called a tapetum lucidum. It reflects light back through the retina allowing the animal to see more clearly in dim light. The tapetum lucidum is what makes animals' eyes light up if you shine a torch at them in the dark.

The opening in the middle of the iris is called the pupil. It allows light to enter the eye.

As light travels through the cornea, a transparent layer at the front of the eye, it is bent or refracted.

The lens is a clear, firm structure that can change shape to bring different areas of vision into focus.

The sclera is the white part of the eye. It is a tough covering that keeps the eyeball safe.

The optic nerve transfers electrical signals from the retina to the brain.

Domestic cat

THIRD EYELID

Many animals have an extra see-through layer between the eyelid and the surface of the eye, called a third eyelid or nictitating membrane. Its job is to protect the eye and keep it moist by spreading tears over the surface.

EYES

Eyes allow animals to see. In their simplest form, eyes are light sensors that determine light from dark, but in many animals, they are complex **organs** that provide a detailed picture of the world around them. When light enters a cat's eye, it passes through many layers before hitting a light-sensitive sheet, called the retina, where special **cells** convert the information into electrical signals, which are sent to the **brain**.

Dilated pupil **Normal pupil** **Constricted pupil**

PUPIL DILATION

A cat's pupil can change size and shape for many different reasons. In low light conditions its pupils expand, or dilate, to allow more light to enter the eye, while in bright light, its pupils contract, or constrict. Cat pupils also dilate when cats are hunting, afraid, or in pain.

BINOCULAR VISION

Like many **predators**, cats have eyes in the front of their head, facing forwards. This means that both eyes can look at the same thing at the same time. We call this binocular vision and it allows animals to judge distances much more accurately.

SENSES

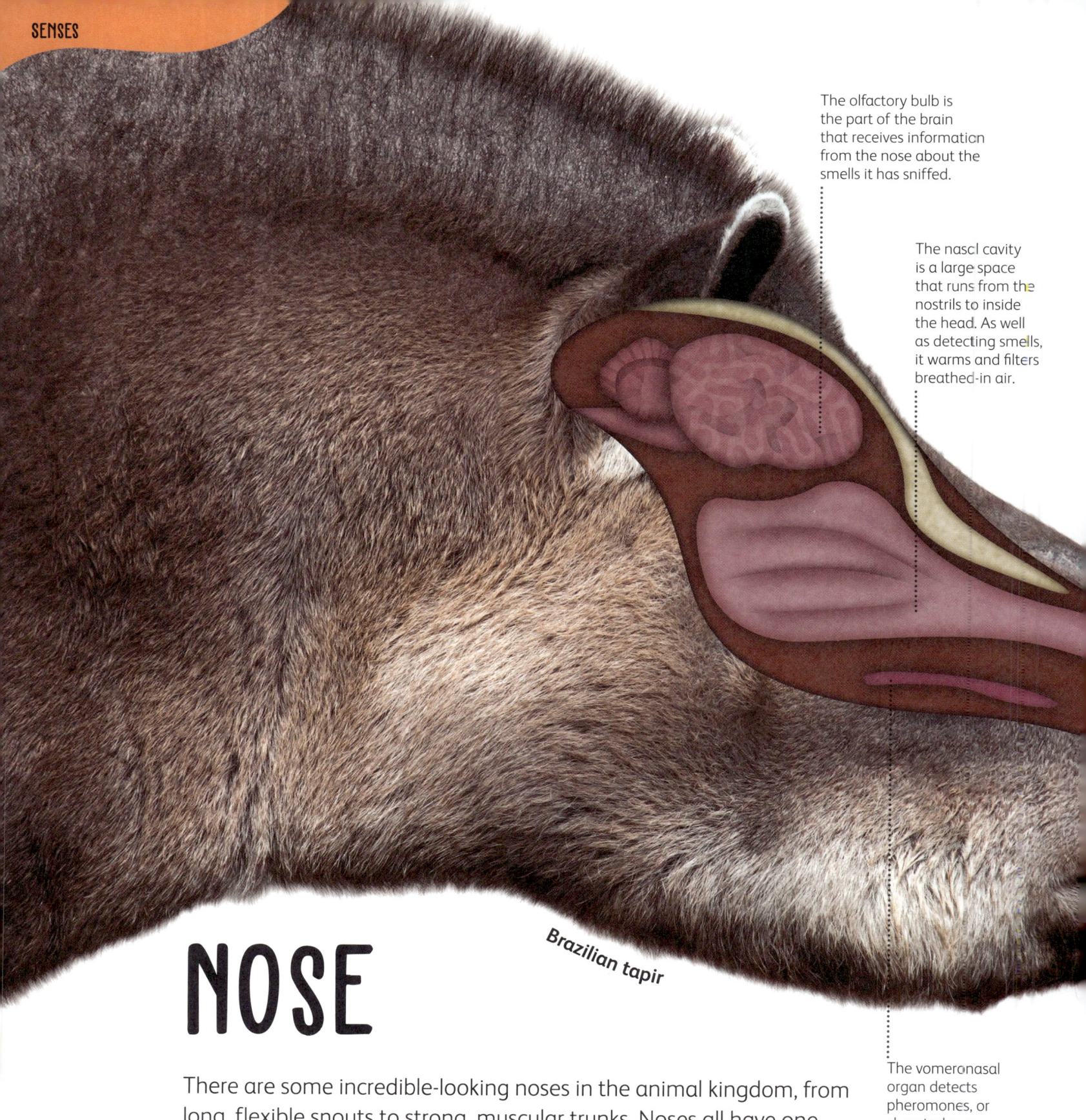

The olfactory bulb is the part of the brain that receives information from the nose about the smells it has sniffed.

The nasal cavity is a large space that runs from the nostrils to inside the head. As well as detecting smells, it warms and filters breathed-in air.

Brazilian tapir

The vomeronasal organ detects pheromones, or chemical messages, from other tapirs.

NOSE

There are some incredible-looking noses in the animal kingdom, from long, flexible snouts to strong, muscular trunks. Noses all have one crucial function though – smelling! When animals sniff the air, tiny floating particles travel up through the nostrils and are trapped in patches of mucus-covered skin at the back of the nasal cavity. Here, special receptors send information about the particles to the **brain**.

MINIATURE TRUNK

The tapir's nose and upper lip are extended into a long, flexible snout called a proboscis, which is a bit like a short elephant trunk. It uses its proboscis for plucking leaves and shoots from trees, as well as for snuffling along the forest floor and sniffing out mates.

FLEHMEN RESPONSE

Some **mammals** signal that they are ready to **mate** by giving off special **chemicals** called pheromones. The pheromones are detected by an **organ** known as the vomeronasal organ, which is located at the base of the nasal cavity. To direct smells to this organ, animals may curl up their lips in a movement called the flehmen response.

The surface of the nasal cavity is covered in **mucus**, which traps germs and prevents them from entering the **lungs**.

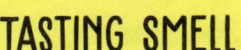

The external openings of the nose are called nostrils. Most animals have two.

TASTING SMELL

For snakes, the main smelling organ is the tongue rather than the nose. Though they can sniff with their nostrils, if they find something that smells particularly interesting they flick out their tongue and waft scent particles into their mouths, where they travel to the vomeronasal organs for identification.

EARS

Sound is produced when a wave of vibrations passes through the air. **Vertebrates** detect these sound waves using special receptors, which are usually found in the ear. The range of sounds that animals can hear is extremely variable — some animals are better at hearing high-pitched sounds, others specialize in noises that are very deep. Owls use their excellent sense of hearing to hunt, and they are extremely sensitive to quiet sounds.

The boreal owl has asymmetrical ears, with the right ear higher than the left.

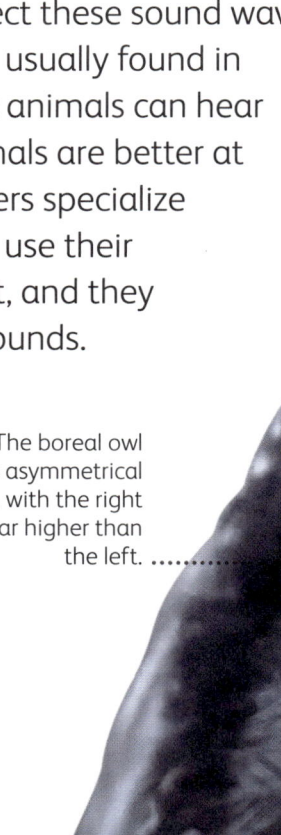

FACIAL DISC

If you look at the face of an owl, it looks like it is made of two dishes placed side by side. These curved, feathery circles make up the facial disc, which acts as a funnel to guide sounds down into the ears.

DEADLY HUNTERS

Most owls hunt at night, using their superb hearing to pinpoint their prey in the dark. To help them find the exact spot where their food is hiding, some have an extra **adaptation** – one of their ears is slightly lower than the other. This makes it easier for them to determine the direction a sound came from.

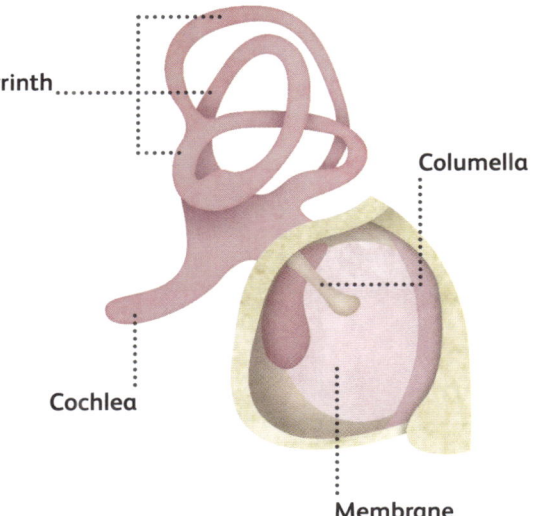

Labyrinth
Columella
Cochlea
Membrane

Boreal owl

There are openings in the **skull** through which the ear canals pass.

INNER EAR

The opening to each of a **bird's** ears is usually covered by feathers. Inside, a tube called the ear canal leads to a thin **membrane**, known as the eardrum. Unlike in **mammals**, vibrations are transmitted by a single small **bone**, called the columella. This goes from the eardrum to the cochlea, which sends sound signals to the **brain**. A structure called the labyrinth helps with balance.

LARGE PINNA

In mammals, the outer ear – the bit we can see from the outside – is called the pinna. Like the owl's facial disc, the pinna acts as a funnel, guiding sounds down through the ear canal to the eardrum. For many species, such as the long-eared jerboa, the pinna also plays a role in keeping cool and communicating.

SENSES

WHISKERS

Animals use their skin and hairs to feel the world around them. While skin can sense changes in temperature, or pain and itches, hairs are best at sensing movement. Most **mammals** have some special thick hairs that are modified specifically for this purpose and are much more sensitive. These hairs are often found around the eyes and muzzle, and we call them whiskers.

Whiskers above the eyes help to protect them by sensing any objects that are near the head.

Whiskers are usually found all around the snout.

HUNTING UNDERWATER

Otters use their highly sensitive whiskers to explore the watery **environments** in which they hunt. As they often search for their food at night, sometimes in murky water, these special hairs allow them to locate **fish** through the vibrations they make in the water, even when visibility is poor.

Whiskers can be very long.

Otters have dense fur, which helps keep them warm and dry underwater.

Asian small-clawed otter

WHISKING

Many mammals, particularly rodents, constantly move their whiskers, in a motion called whisking. This particularly happens when they are moving or exploring and involves sweeping their whiskers very quickly from side to side — sometimes moving them up to 25 times a second!

- Whisker
- Skin
- Nerve
- Follicle
- Pressure and movement sensors
- Nerve

WHISKER STRUCTURE

Like all hairs, whiskers are made of **keratin** and grow from pockets in the skin, called follicles. Unlike normal hairs, though, whiskers grow from follicles found much deeper in the skin. The follicles also contain extra sensors that are sensitive to changes in pressure and movement, so when a whisker is moved even slightly, a signal is sent to the **brain**.

EXTRA SENSES

Animals don't all experience the world in the same way. Some can see, hear, smell, taste, and feel things that others cannot. In addition to the five main senses, some animals also have extra special senses that allow them to perceive their surroundings in a different way entirely.

ELECTRORECEPTION
The bill of the platypus is covered in small electrical sensors, which pick up the tiny electrical signals created by the **muscle** contractions of its **prey**. This allows it to find its food in murky water.

MAGNETORECEPTION
Many **birds**, especially those that **migrate** long distances, can sense the Earth's magnetic field. They use it like a compass, helping them to find their way.

INFRARED SENSE
Some snakes, such as pit vipers, have heat vision! Pits on their face are filled with sensors that can detect changes in temperature, helping them to find their warm-bodied prey and aim their strikes.

LIFE SIZE

ECHOLOCATION

Some dolphins and whales, such as the beluga, navigate and find their prey using **echolocation**. They send out sounds into the water and listen for the returning echoes to build up a picture of their surroundings.

ULTRAVIOLET SENSE

Birds can see a much wider range of colours than humans can, including many in the ultraviolet range. Birds that look quite dull to human eyes may have very beautiful UV patterns!

LATERAL LINES

Fish have a sensory system called the lateral line system, which allows them to feel vibrations or movement in the water around them. The **pores** that open to the lateral line system can be seen running down either side of the fish.

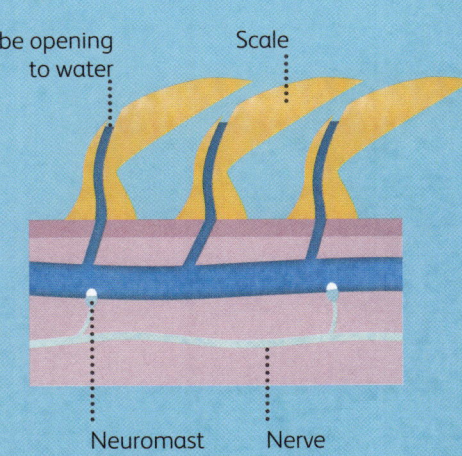

Tube opening to water — Scale — Neuromast — Nerve

Sensing pressure

Each lateral line is made of a long tube just under the fish's scales with pores connecting to the surrounding water. In the tubes are sensors called neuromasts, which have tiny hairs attached. When the hairs are bent to one side by changes in water pressure, they send a signal along a nerve to the **brain**.

SENSES

INSIDE AN ELEPHANT

This majestic giant of the African savannah has an incredible array of sensory **organs**. Not only are an elephant's ears and trunk impressive to look at, they are brilliant tools too. With excellent hearing and a keen sense of smell, elephants experience the world in a way that is totally unique, relying very little on their eyesight. They also have a special **adaptation** buried in the soles of their feet!

SUPER SMELLERS

Elephants have an extraordinarily good sense of smell, with more smell receptors in their nose than any other animal. A huge portion of their brain is also dedicated to processing scents. Smelling is the elephant's primary way of experiencing the world.

Although elephant eyes are relatively large, their vision is poor compared to many other **mammals**.

Elephants have the largest **brains** of all land animals.

The nasal cavity has two branches that travel all the way down the trunk, running to two nostrils at its end.

Nerves all along the trunk send sensory information to the brain.

MULTIPURPOSE TRUNK

Elephant trunks have many functions; they can be used as snorkels for swimming, hoses for washing, and hands for grabbing. Though they might not seem very dexterous, finger-like projections at the tip of the trunk are gentle and nimble enough to pick up objects as small as a blade of grass!

106

LARGE EARS

An elephant's enormous ears have a number of functions: they help to regulate body temperature by dispersing heat, they can be flapped and spread wide to warn off other elephants, and they act as huge funnels that direct sound down towards the ear canals.

Around 800,000 nerves work together to control an African elephant's enormous trunk.

LISTENING FEET

Despite their huge and magnificent ears, elephants may detect sound through another body part too – their feet! Strictly speaking, they are actually feeling rather than hearing the sound. The vibrations of low-pitched noises travelling through the ground are picked up by receptors in the skin called Pacinian corpuscles.

African elephant

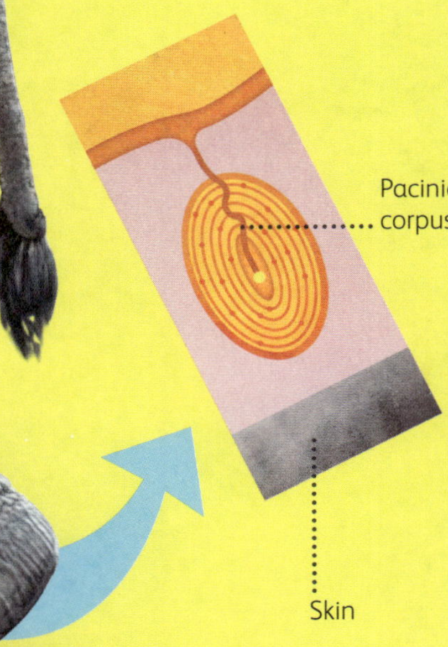

Pacinian corpuscle

Skin

REPRODUCTION

SPERM AND EGGS

The reproductive organs produce special **cells** called gametes, which contain half the information needed to make a new individual. Male gametes are called sperm and female gametes are called eggs. One sperm and egg join together in a process called fertilization to create an embryo, which will grow into a baby.

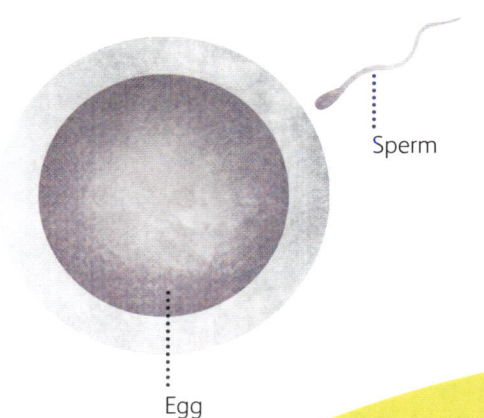

Sperm

Egg

Urethra
The urethra is a tube that runs from the bladder through the penis. Sperm and **urine** exit the body through it.

Testes
Sperm are made in a pair of organs called testes.

Penis
In some species, sperm is transferred to a female's vagina with an organ called the penis.

Male lion

WHAT IS REPRODUCTION?

For a **species** to survive, the animals belonging to it must produce offspring. In many species, this requires a male and a female to **mate**, so their sperm and eggs can join in a process called sexual reproduction. This type of reproduction produces babies that have a mixture of the features of their two parents.

FERTILIZATION

In some species, fertilization occurs inside the body, with the sperm travelling up into the uterus to meet an egg – this is called internal fertilization. In other species, it occurs outside the body. This is the case for most **amphibians** and **fish**, which release their gametes into water at the same time – this is called external fertilization.

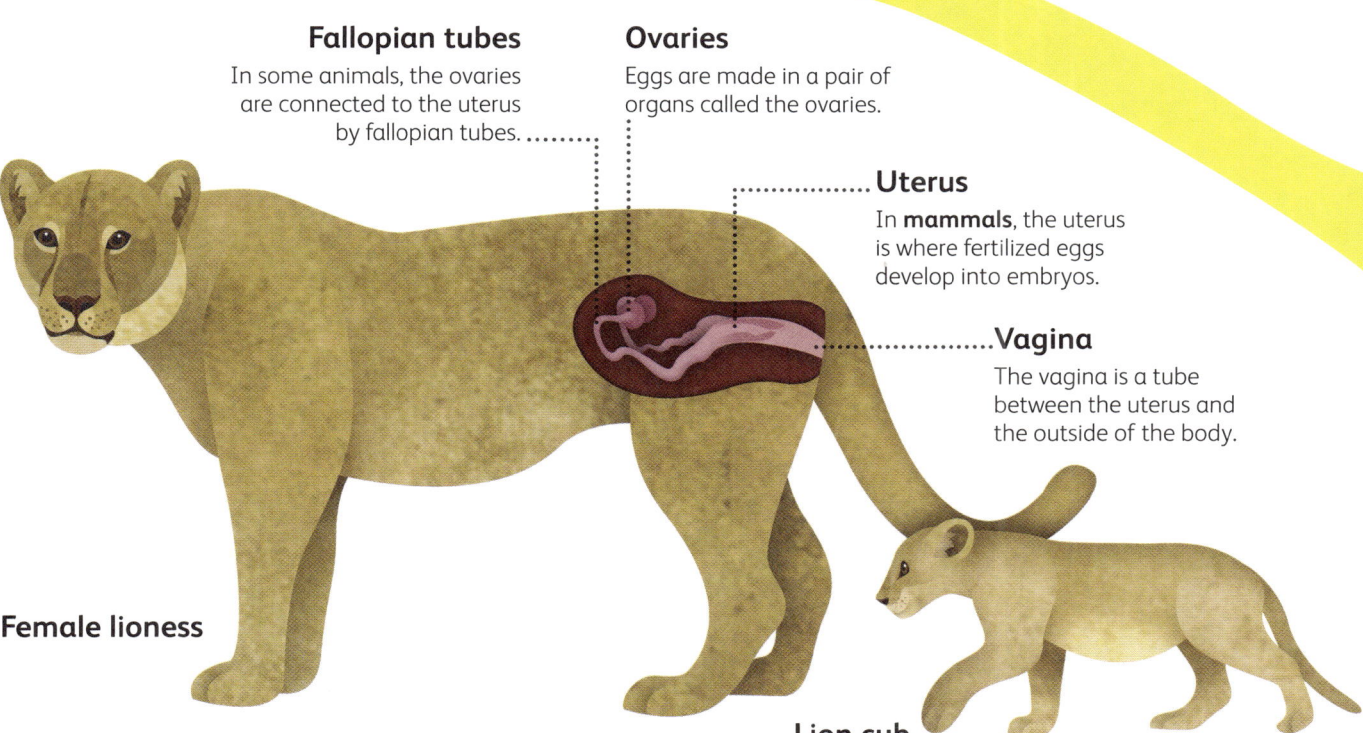

Fallopian tubes
In some animals, the ovaries are connected to the uterus by fallopian tubes.

Ovaries
Eggs are made in a pair of organs called the ovaries.

Uterus
In **mammals**, the uterus is where fertilized eggs develop into embryos.

Vagina
The vagina is a tube between the uterus and the outside of the body.

Female lioness

Lion cub

MALE AND FEMALE

The reproductive system differs between males and females. They have different sets of reproductive **organs**, which produce a variety of **chemicals** called **hormones** that affect their behaviour and body shape in different ways.

Offspring
Some animals lay eggs and others give birth to live young, such as a lion cub. Lionesses care for their babies until they are grown, but other animals leave their offspring to fend for themselves.

CLOACA

Male and female **birds** and **reptiles** have different reproductive organs, but these are hidden away inside the body. From the outside, all that can be seen is an opening called the **cloaca**, where the reproductive, digestive, and urinary systems meet.

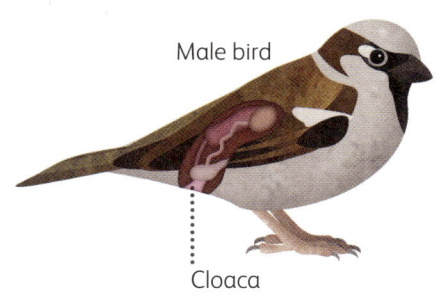

Male bird

Cloaca

Female bird

Cloaca

III

REPRODUCTION

EGGS WITHOUT SHELLS

CHOOSING A MATE

Male toads call loudly to attract the attention of a female. When a female toad chooses her mate, he jumps onto her back and holds on tight in a position known as amplexus. He will remain there until she lays her eggs, so he is ready to fertilize them when she releases them – this can sometimes take months!

Fish and **amphibians** lay soft eggs without shells. Although these eggs are surrounded by a jelly-like protective layer, they will soon dry out if they are away from water for too long, so they must always be laid in a very damp or wet **environment** to keep them safe. Some reptiles lay soft eggs too, but they are surrounded by a **membrane** called an amnion, which prevents them from losing too much moisture.

Toads have two ovaries. During the breeding season, they are full of eggs.

The oviducts are long, coiled tubes down which the eggs pass to the uterus.

A jelly coating is added to the eggs in the oviducts. The jelly swells up when it meets water.

The eggs leave the body through the **cloaca**.

STRINGS OF EGGS

Unlike frogs, which lay their eggs in big clumps, toads lay their eggs in strings. There are two rows of eggs inside each string – one from each ovary. The female European green toad can lay more than 9,000 eggs in one go and its egg strings can reach 7 m (23 ft) in length!

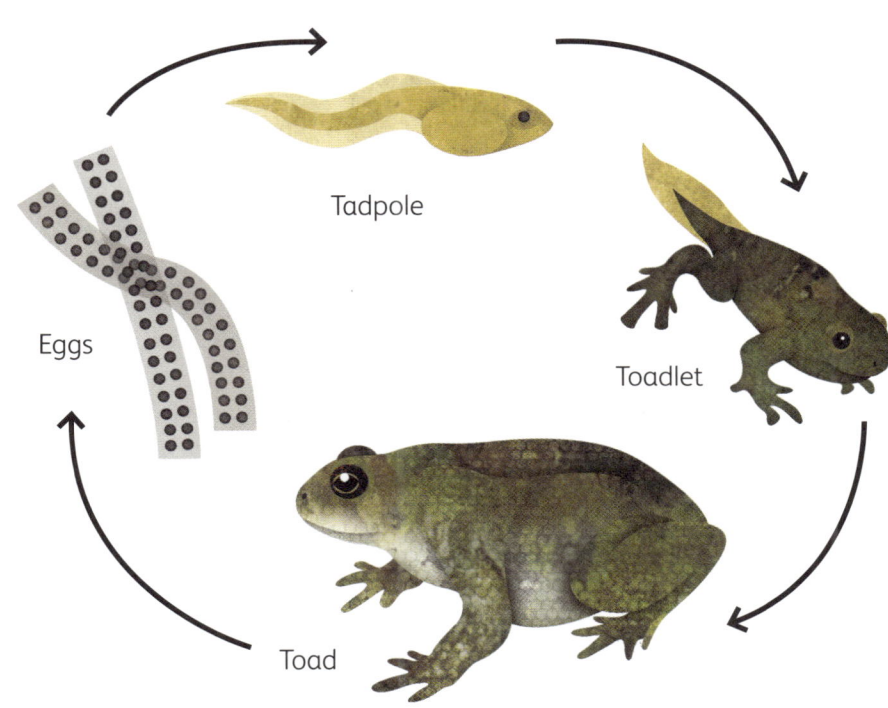

Eggs

Tadpole

Toadlet

Toad

LIFE CYCLE

When toad eggs hatch, limbless tadpoles with long tails emerge. They look nothing like adult toads, but over the next few months they will lose their tail and grow legs, completely changing their body shape. First they become toadlets, and finally adult toads, in a process called metamorphosis.

Female European green toad

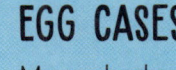

EGG CASES

Many sharks also lay eggs, but these are covered in a tough, leathery egg case. These egg cases often have long, curled tendrils or sticky hairs that attach them to underwater plants. It may take many months for the baby shark to hatch, but when it does it will look like a tiny version of an adult shark.

REPRODUCTION

The eggshell is formed around the albumen by the shell **gland**. Any patterns on the egg are added just before it is laid.

Most birds have only one ovary but kiwis have two. This is where the yellow yolks are made.

The egg exits the body via the **cloaca**.

Each ovary has an oviduct — a tube which delivers the egg to the cloaca. The albumen is added around the yolk here.

EGGS WITH SHELLS

Birds, some **reptiles**, and a handful of **mammals** lay eggs enclosed in hard shells to stop them from drying out. Bird eggs have shells made from **calcium**, the same **mineral** found in **bone**, and they are surprisingly difficult to break. It's impossible to see them, but tiny **pores** allow **oxygen** to pass through the eggshell so the developing chick inside can grow. Kiwis lay some of the biggest eggs in the world, but usually only one at a time.

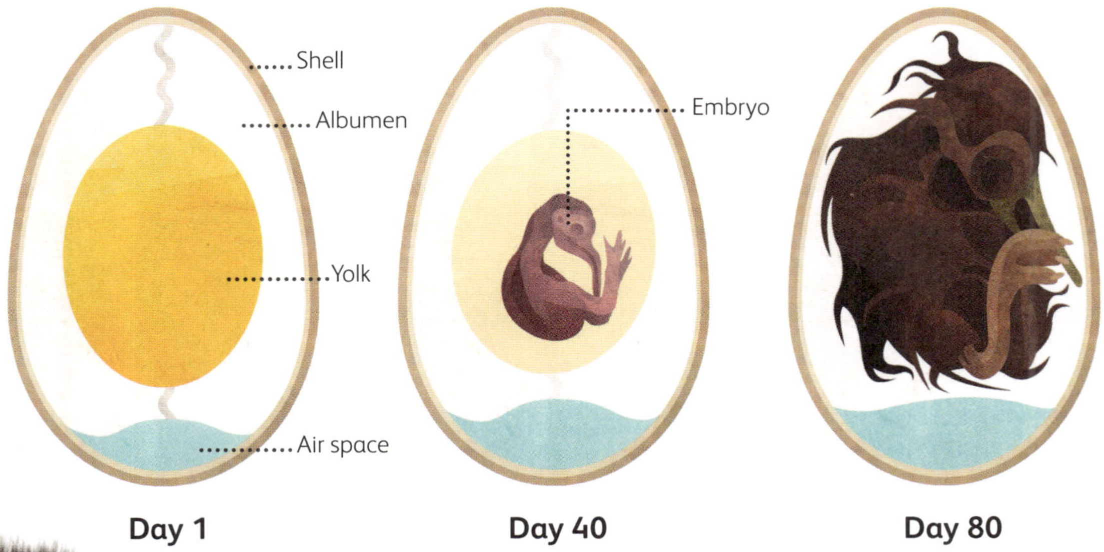

Day 1 · Day 40 · Day 80

- Shell
- Albumen
- Yolk
- Air space
- Embryo

INSIDE THE EGG

Inside the egg, the developing chick has everything it needs to grow. In the centre is the yolk, full of **fats** and **nutrients**, and around this is the albumen, or egg white, which contains water. Separating the albumen from the shell are two tough **membranes**, which protect the developing chick from **bacteria**. Brown kiwis incubate their egg for a long time – around 80 days!

Brown kiwi

ENORMOUS EGG

Kiwi eggs are enormous! The **skeleton** above shows how much space the egg takes up in the mother's body. Each egg can be a whopping 20 per cent of its mother's weight. The eggs contain so much nutritious yolk that kiwi chicks don't need to eat for the first two weeks after hatching.

REPTILE EGGS

Most reptiles lay soft-shelled eggs but crocodiles, alligators, and some tortoises lay hard-shelled eggs like birds. Just like bird eggs, reptile eggs contain a yolk which feeds the baby as it grows. A baby tortoise's carapace and plastron are soft when it hatches so it can fit inside the egg.

REPRODUCTION

MILK
All species of mammal produce milk, a rich liquid made in the female's mammary **glands**, to feed their babies. A special milk called colostrum is made in the first days after birth which has extra **nutrients**. When young, mammal offspring don't need any other food than milk – a lamb will drink only milk for up to six months.

Female sheep

FEMALE ANATOMY

The offspring of most **mammals** develop inside their mother's body, in an **organ** called the uterus. They are supported by another amazing organ called the placenta, which allows the mother to give food and **oxygen** to, and remove waste products from, them. The uterus must be very stretchy to allow the developing babies to grow. After birth, female mammals also produce milk, which their young rely on for food.

SLOW GROWING
The amount of time a baby spends growing inside its mother's body is called gestation. It varies wildly from **species** to species. The shortest gestation of any mammal is in the Virginia opossum, which gives birth after just 12 days. The longest is in the African elephant, which can carry its baby for up to 22 months.

REPRODUCTIVE SYSTEM

In most mammals, the female reproductive system is made up of the same set of **organs**. Eggs are produced in the two ovaries. These are connected by fallopian tubes to the uterus, where young develop. In animals such as sheep, the uterus has two pointed ends called horns. Another tube, called the vagina, leads from the uterus to the outside of the body.

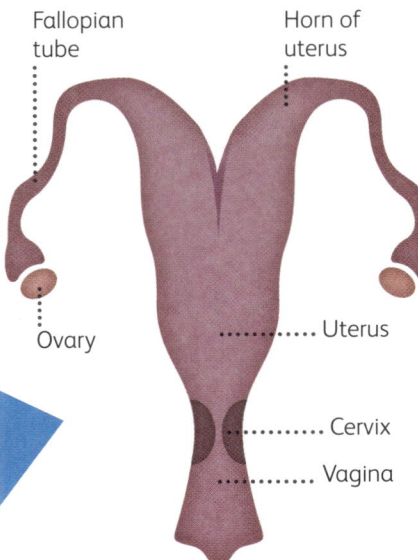

Fallopian tube

Horn of uterus

Ovary

Uterus

Cervix

Vagina

The uterus is where the baby grows inside the mother.

The developing baby is held in a fluid-filled bag called the amniotic sac.

Babies pass down the vagina, also called the birth canal, when being born.

The cervix is the muscular neck of the uterus, which stays closed until a baby is ready to be born.

The placenta acts like a bridge between the **blood** supply of the mother and the baby.

In sheep, the uterus connects with the placenta through a round structure called a caruncle.

The umbilical cord connects the lamb to the placenta.

The mammary glands of sheep and cows are known as udders.

REPRODUCTION

Brown rats don't have a breeding season, and can produce young all year round.

Male brown rat

LARGER MALES

In some animals, males and females of the same **species** vary in size, this is known as sexual dimorphism. In most sexually dimorphic animals, females are bigger, but for **mammals**, it can be the reverse – males are sometimes the larger sex. This is likely to give them an advantage when competing for a **mate**.

MALE ANATOMY

The main function of the male reproductive system is to produce sperm and deliver them to the site of the female's eggs to fertilize them. As well as producing sperm, the male reproductive **organs** produce testosterone, a **hormone** that plays an important role in the development of male features and behaviours. The urinary system is also closely linked to the reproductive system in males.

BACULUM

Some animals, including raccoons, have a bone inside their penis called a baculum. Not all species have a baculum, but in those that do, it comes in a great array of shapes and sizes. It can help males to mate successfully.

Bladder
Seminal vesicle
Vas deferens
Testis
Urethra
Penis
Prostate gland
Testis

REPRODUCTIVE SYSTEM

The male reproductive system of mammals is made up of several organs: the testes are where sperm and testosterone are made; the vas deferens is a tube that carries sperm from each testis to the prostate **gland**; the prostate and seminal vesicles help make semen, the fluid in which sperm is carried; and the penis, which is held to, or inserted into, a female's vagina during **mating** to deliver sperm to fertilize her eggs.

The vas deferens, or sperm ducts, are muscular tubes that connect the testes with the penis.

In mammals, the penis can usually be seen outside the animal's body.

The urethra is the tube that runs through the penis, carrying both **urine** and semen.

The testes are held in a sac of skin called the scrotum.

REPRODUCTION

TRAGOPAN

To impress females, male Temminck's tragopans puff up their feathers, spread their wings, bob their tails, and make distinctive calls. They also reveal a brightly patterned bib under their chin, and two blue horns.

ORNAMENTS

For many animals, being in with a chance of **mating** means impressing the opposite sex. Some animals take this challenge very seriously and have some highly impressive features to help them in their task. When an animal has a body part whose function is the way it looks, we call it an ornament.

SHOWING OFF

Sometimes the extent of an ornament is only revealed during courtship displays, when an animal tries to draw attention to itself. The male tragopan's blue horns and bib are only revealed when he is displaying.

LIFE SIZE

PEAFOWL

Male peafowl, known as peacocks, have enormous and magnificent tail feathers. When displaying, they fan the feathers out, and vibrate and rattle them to attract the attention of nearby female peafowl, known as peahens.

PROBOSCIS MONKEY

Male proboscis monkeys have large, pendulous noses, which are thought to be attractive to female monkeys. A male's nose can reach 10 cm (4 in) long!

INDIAN BULLFROG

Male Indian bullfrogs go from green to bright yellow when it is time to mate. The sac under their throat, which is used for calling, turns deep blue!

SOCKEYE SALMON

During the breeding season, adult male sockeye salmon develop humped backs and hooked **jaws** filled with tiny teeth. Both males and females also turn bright pink.

RED PHALAROPE

It is rare for female birds to be more brightly coloured than males, but female red phalaropes are! When they are ready to mate, they turn a bright and vivid red.

REPRODUCTION

INSIDE A KANGAROO

Kangaroos belong to a group of **mammals** called marsupials. This means the females give birth to tiny babies after a very short gestation period and care for them in a pouch until they are big enough to fend for themselves. Each kangaroo mother can care for two babies, known as joeys, of different ages, and hold another fertilized egg in reserve. Unsurprisingly, the marsupial reproductive system looks very different from that of other mammals.

SAFE POUCH
Red kangaroo joeys can spend up to eight months in their mother's pouch. During the last few months they will leave for short periods, returning if they feel threatened or to suckle on their mother's milk. These joeys are called "young at foot".

Only female marsupials have pouches.

DIFFERENT MILKS
A kangaroo mother can feed two joeys of different ages at the same time. Amazingly, she produces a different kind of milk for each baby, depending on what it needs! To make sure it always gets the correct milk, each joey only feeds from one particular nipple.

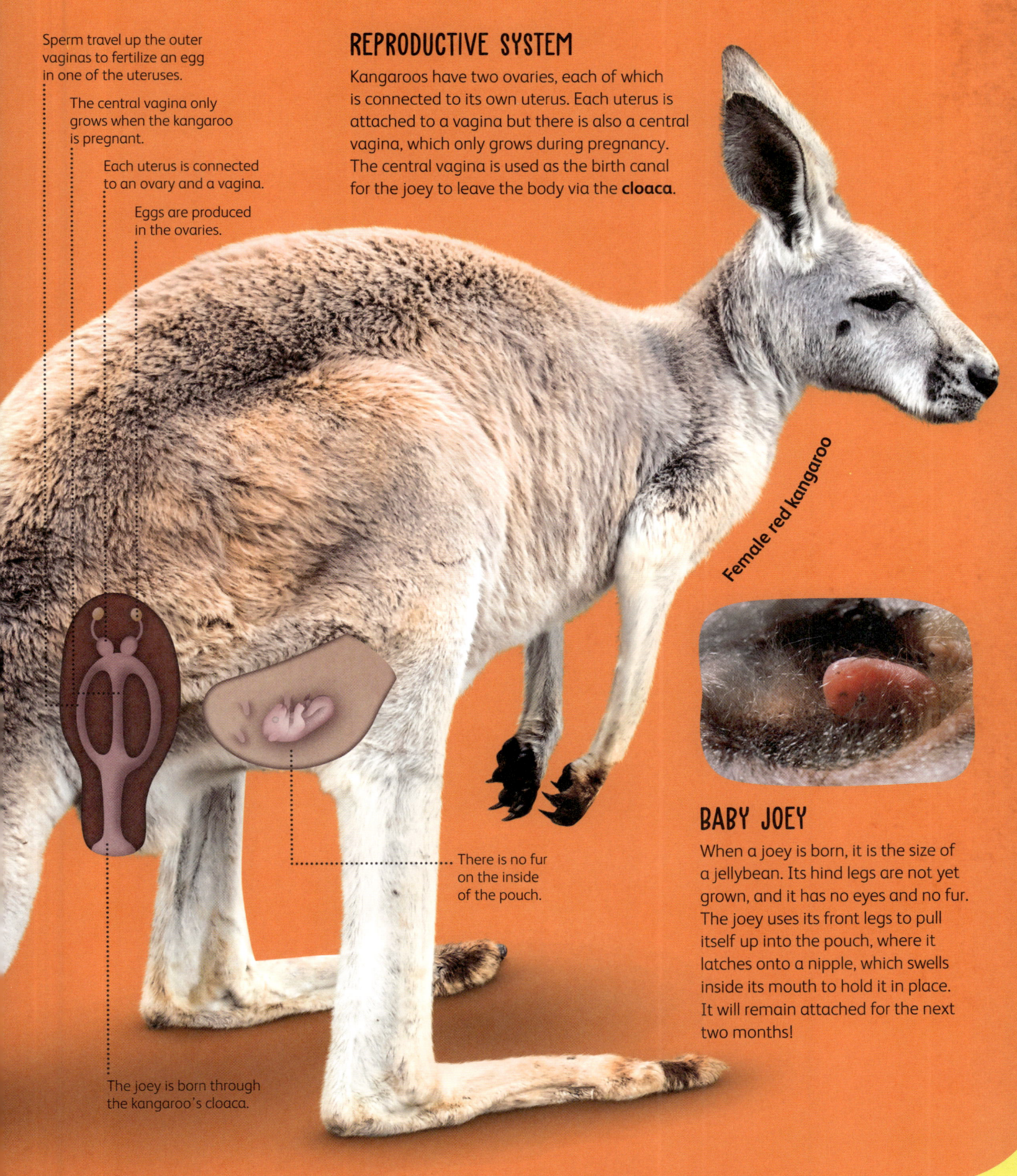

Sperm travel up the outer vaginas to fertilize an egg in one of the uteruses.

The central vagina only grows when the kangaroo is pregnant.

Each uterus is connected to an ovary and a vagina.

Eggs are produced in the ovaries.

REPRODUCTIVE SYSTEM

Kangaroos have two ovaries, each of which is connected to its own uterus. Each uterus is attached to a vagina but there is also a central vagina, which only grows during pregnancy. The central vagina is used as the birth canal for the joey to leave the body via the **cloaca**.

Female red kangaroo

There is no fur on the inside of the pouch.

The joey is born through the kangaroo's cloaca.

BABY JOEY

When a joey is born, it is the size of a jellybean. Its hind legs are not yet grown, and it has no eyes and no fur. The joey uses its front legs to pull itself up into the pouch, where it latches onto a nipple, which swells inside its mouth to hold it in place. It will remain attached for the next two months!

INTEGUMENT

SKIN

Skin is a flexible layer that helps stop animals from losing water and which protects them from germs and other hazards. Hippos live out in the open in Africa, where the strong sun could give them sunburn. To protect themselves, they release a red, oily sunscreen from **glands** beneath their skin!

Epidermis
The outermost layer of the skin, the epidermis, is constantly replaced as old dead skin **cells** fall off.

Dermis
The middle layer of the skin, the dermis, contains nerves, **blood vessels**, follicles, and glands.

Hypodermis
The innermost layer, the hypodermis, anchors the skin to the **muscle** below. It is mostly made of **fat** and **connective tissue**.

Sweat pore
Sweat is produced in sweat glands and released through small holes in the skin, called **pores**.

Nerves
The skin contains many nerves, which allows it to sense the outside world.

Hair
Hairs grow from pockets in the dermis called follicles. Tiny hair muscles allow the hairs to rise up in cold weather and trap warm air.

Blood vessels
As well as delivering **oxygen** and **nutrients** to the skin, blood vessels in the dermis help to regulate temperature.

WHAT IS INTEGUMENT?

All **vertebrates** have a protective covering called skin. In many animals, the skin is the biggest **organ** in the whole body! It is made up of three layers: the epidermis, the dermis, and the hypodermis. A range of different coverings are found growing from the skin, including hair, feathers, scales, and plates. Together, these body parts make up the integument – a barrier between the inside of the animal and the outside world.

OUTER COVERINGS

In addition to skin, most animals have some extra coverings that grow from the skin to help protect and insulate themselves. Some of these coverings, such as a **bird's** feathers, even help them move around.

KERATIN

Hair, nails, hooves, claws, horns, feathers, spines, and the beaks of birds and turtles are all made of an incredible material called **keratin**. Keratin is a type of tough **protein** that is also found in the outer layer of the skin.

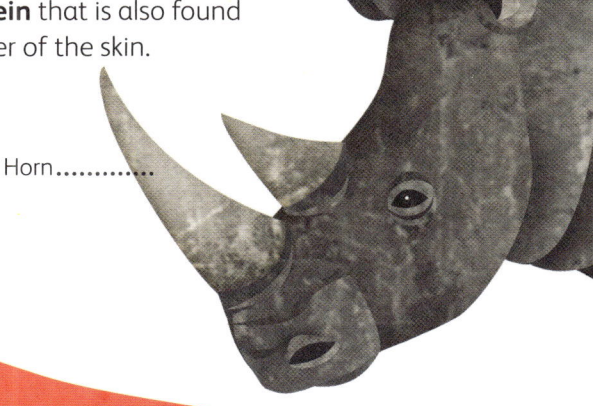

Horn............

Hair
Hair grows on the bodies of **mammals**. It can vary in length, thickness, colour, and patterning depending on the animal's **environment**. Whales only have a few strands of hair, while llamas have a thick fur coat.

Plates
Pangolins are the only mammals whose bodies are covered in plates, a type of hard scale. These overlapping structures protect the pangolin from predators, especially when it rolls into a ball. In some animals, plates are called scutes.

Scales
Found on the bodies of **fish** and **reptiles**, scales are tough sheets that provide protection. Fish scales are individually embedded in the skin, but reptile scales form a continuous smooth surface.

Feathers
Feathers are unique to birds. They come in many different shapes and have many uses. Fluffy feathers help with insulation, while longer, flat feathers help with flying, waterproofing, or even swimming.

INTEGUMENT

BLUBBER

Some marine animals, including walruses, have a thick layer of **fat** in their skin to keep them warm, we call this **blubber**. Blubber is a brilliant insulator and is really effective at preventing body heat from being lost to the surrounding water. It also helps animals to stay afloat, protects them against predators, and acts as an extra store of energy for use when food is scarce.

KEEPING AFLOAT

Fat is less dense than water, meaning it floats. This helps walruses to move around their aquatic environment without using up too much energy. Marine **mammals** can also control whether they float or sink by emptying and filling their **lungs** with air.

On its body, the blubber of a walrus can be 15 cm (6 in) thick.

Walruses can curl their rear flippers under their bodies to walk on all fours on land.

NURSING MOTHER

It takes a lot of energy for walrus mothers to produce milk for their nursing young. Not only that, they must spend a lot of the time they would usually use to search for food caring for their calves. During this time, they rely on the energy stores in their blubber.

Walrus skin is often scarred from fights with other walruses or attempted attacks by predators.

LAYER OF FAT

The fat that makes up blubber is found deep in the skin. Marine animals move around a lot, so they must be able to adapt to a range of different temperatures. When things get particularly cold, they can narrow the **blood vessels** inside their blubber to reduce the amount of heat lost through it.

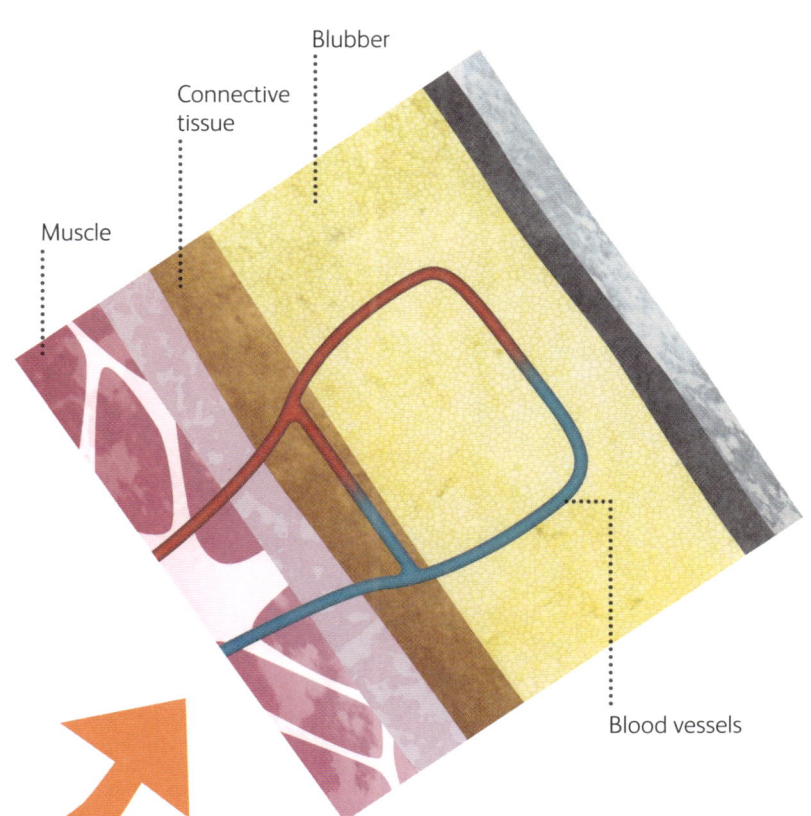

Blubber
Connective tissue
Muscle
Blood vessels

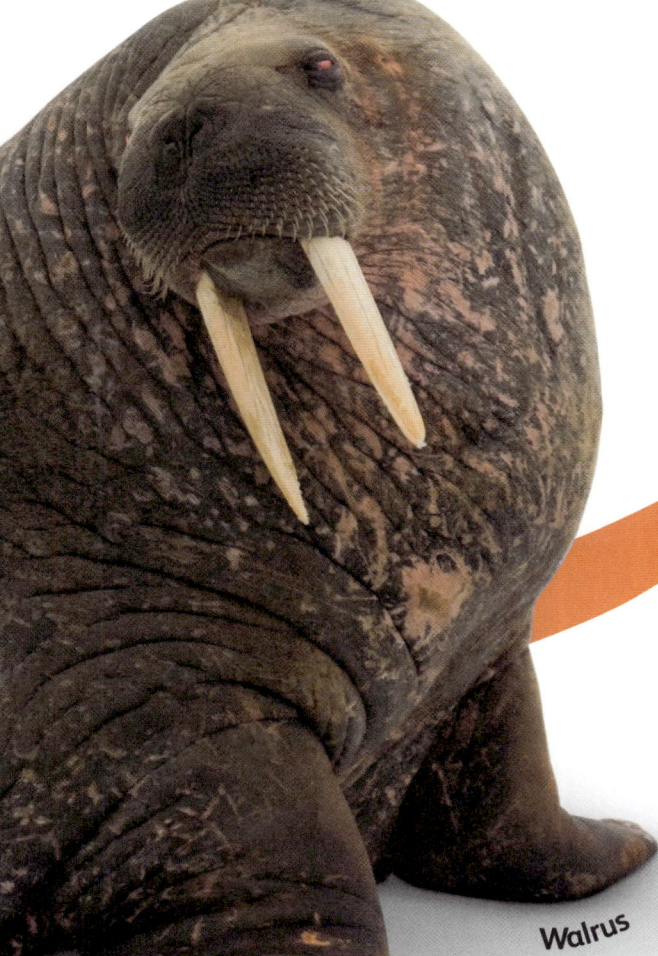

Walrus

The flippers are the only parts of the body that are not covered in a thick layer of blubber.

BROWN FAT

There is a special kind of fat, called brown fat, which works like a built-in heater, creating heat energy from food. Animals that hibernate typically have a lot more brown fat than other animals as it keeps them warm while they sleep. Younger animals also tend to have more brown fat than adults.

129

INTEGUMENT

SKIN

Vertebrate animals are wrapped in a flexible, protective shield called skin. As well as keeping them safe from harm, skin helps animals to regulate their body temperature, communicate, and feel things through touch. **Amphibians**, such as frogs, have very special, extra-thin skin, that is not only used for protection but for breathing too. Some frogs can also release dangerous poisons through their skin to protect them from **predators**.

WRINKLY SKIN

Lake Titicaca frogs live underwater at high altitudes in South America. They have large amounts of very loose excess skin, which falls in baggy folds over their bodies. The extra skin acts like **gills**, providing a larger surface area for oxygen to pass across.

Poison dart frogs have brightly-coloured skin to warn predators not to eat them because of the poisons they produce.

APOSEMATISM

Animals are colourful for all sorts of reasons. Often their colours and patterns are meant to hide them from view, but sometimes they are designed to be seen! Many animals that are poisonous advertise their toxicity through bold, bright colours, particularly in shades of yellow, orange, and red. This is called aposematism.

SLIME NURSERY

Frogs spend most of their time in damp **environments**, but to make sure their skin never dries out they can also produce their own sticky **mucus** from specialized mucous **glands**. Some poison dart frogs also use their sticky skin to carry their tadpoles to tiny pools of water in plants high up in trees, where they may develop safely.

Slimy skin is difficult to grab hold of and can help frogs to escape from predators.

Blue poison dart frog

The mucus produced by frogs has been found to kill some bacteria and fungi.

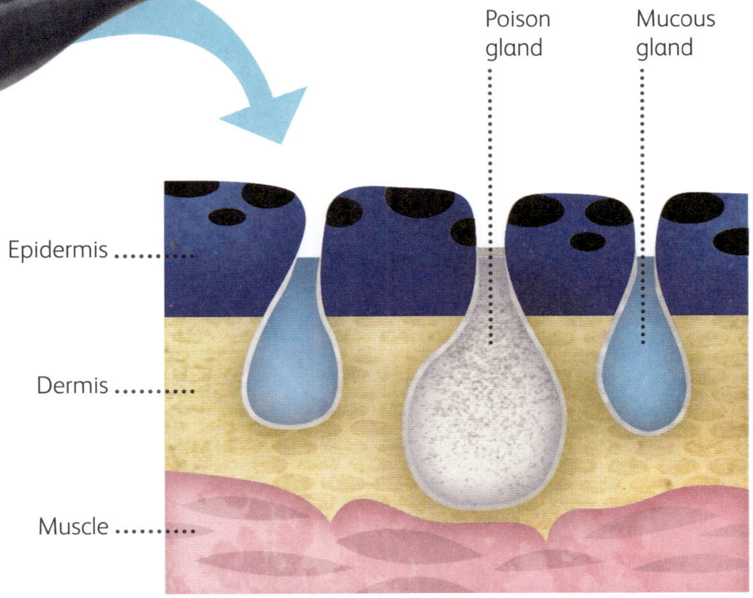

Poison gland • Mucous gland

Epidermis
Dermis
Muscle

SLIMY SKIN

It is really important that frog skin never dries out because it must be wet for gases such as **oxygen** to move through it. As well as mucous glands to keep their skin moist, many frogs have poison glands, which produce **toxic chemicals** that make them unappealing or even dangerous to predators.

INTEGUMENT

SCALES

Reptiles have the perfect protection from wear and tear, and germs – scales! These small, dry plates are made of **keratin**, and a reptile may have thousands of them. Scales are formed from the thick outer layer of the skin, known as the epidermis. They are watertight and can be very brightly coloured. Unlike **fish** scales, which are attached individually, reptile scales are all connected.

SHED SKIN

As they grow, reptiles shed their old skin to replace it with a new fresh and healthy one. Snakes and some lizards will shed their skin in one complete piece, including the clear scales that cover their eyes! Tortoises and the other lizards shed their skin in several pieces.

Tokay geckos have a pattern of orange and turquoise scales.

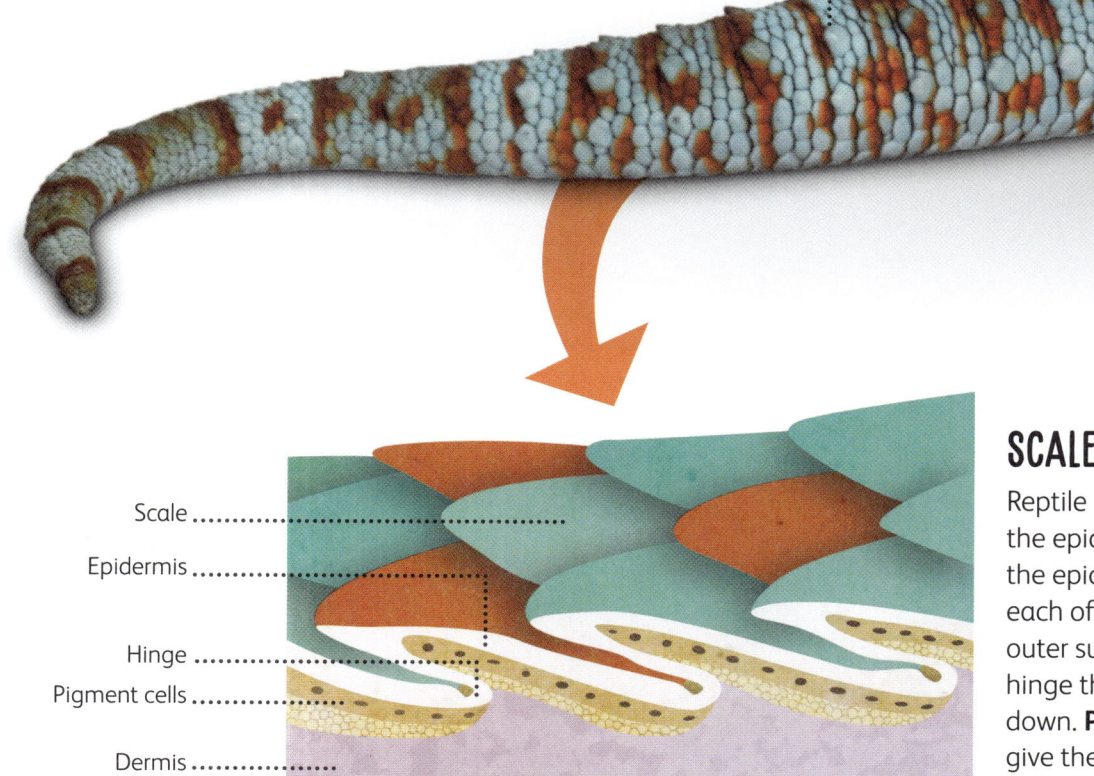

Scale
Epidermis
Hinge
Pigment cells
Dermis

SCALE STRUCTURE

Reptile skin is made of two layers: the epidermis and the dermis. Folds in the epidermis form what we call scales, each of which has three parts – a hard outer surface, a soft inner core, and a hinge that lets the scale move up and down. **Pigment cells** in both layers give the scales their incredible colours.

OSTEODERMS

Some animals have bony plates called osteoderms embedded within their skin for extra protection and support. Osteoderms were found in many extinct reptiles and **amphibians**. They are still present in some living reptiles, including skinks and crocodiles, but are only found in one **mammal** – the armadillo.

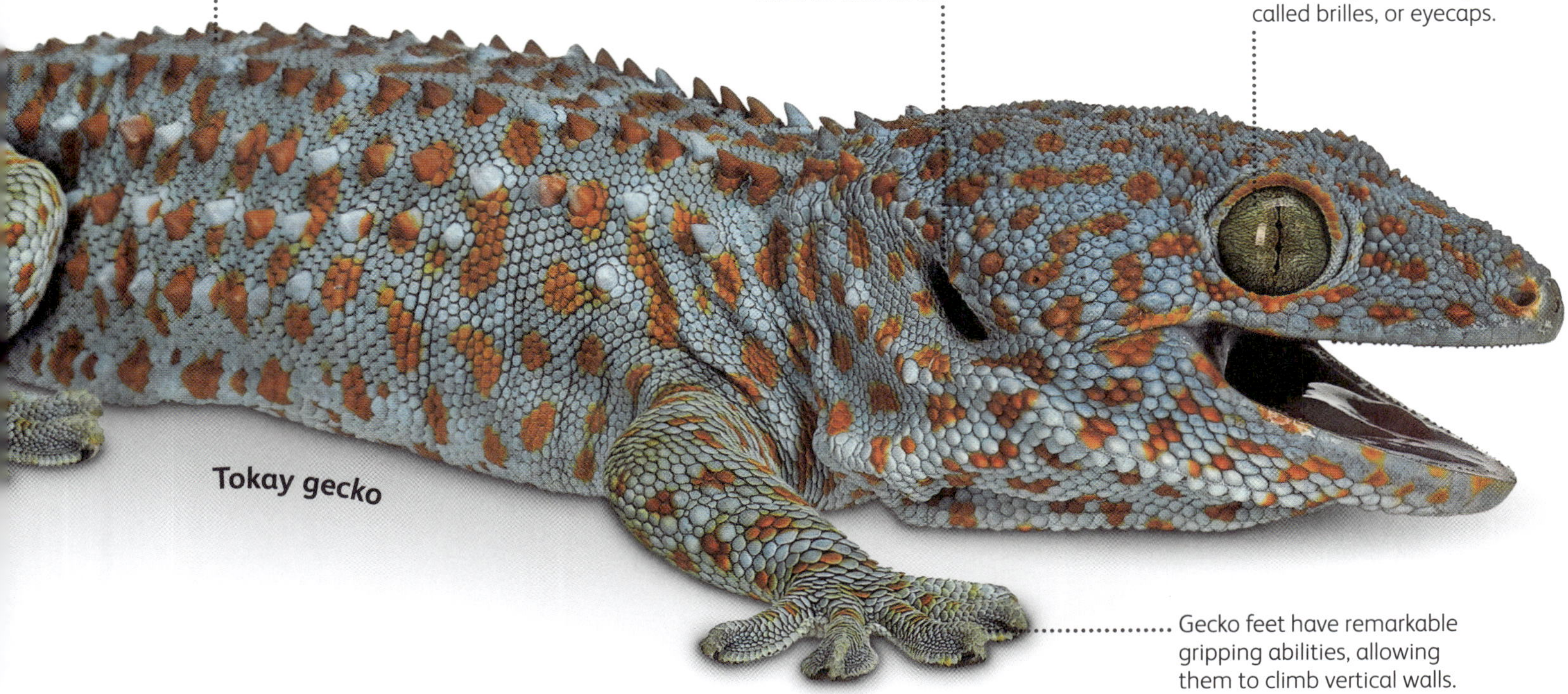

Some lizards have scales that are modified into sharp spines for protection.

Some reptiles have simple ears that don't have an external earlobe. Snakes have no ears at all!

Geckos and snakes don't have eyelids. Instead, their eyes are covered by see-through scales called brilles, or eyecaps.

Tokay gecko

Gecko feet have remarkable gripping abilities, allowing them to climb vertical walls.

GRIPPING FEET

Geckos are brilliant at scuttling up walls and even hanging from ceilings by their feet. This is made possible by tiny, flexible hairlike structures on the scales of their toes, called setae. Each of these setae is in turn covered by even smaller hairs, which are held to the wall by minuscule electrical charges.

INTEGUMENT

The alula is a small collection of up to five wing feathers that helps a bird land and take off.

Coverts are the contour feathers of the wing. They cover the area where the wing feathers attach to the **bone** of the wing.

FLUFFY CHICKS

When they hatch, most birds are covered in soft, fluffy feathers called down feathers. These feathers are brilliant at trapping heat next to the skin, keeping the chicks toasty warm while their parents are away searching for food. Many adult birds have an inner layer of down feathers to keep them warm too.

Bald eagles can have a wingspan of more than 2 m (6 ft).

FEATHERS

The bodies of **birds** are covered in highly-specialized structures called feathers. Birds are the only living animals to have them. Feathers are light and strong, and come in lots of different shapes, sizes, and colours, depending on their function — they can be used for insulation, communication, waterproofing, flying, and sensing. Like the scales of **reptiles** and the hair of **mammals**, feathers are made of **keratin**.

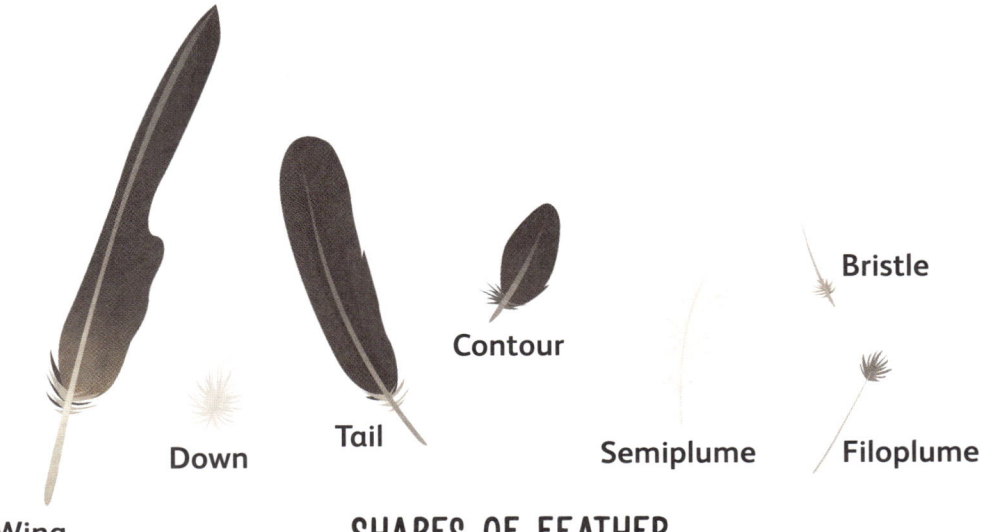

SHAPES OF FEATHER

There are many different shapes of feather for many different jobs: contour feathers give the body its streamlined shape; wing feathers are used for flying; tail feathers are usually used for steering; down and semiplume feathers are for insulation; and bristles and filoplumes act like whiskers, sensing touch.

PREENING

Feathers are very important, so must be kept in top condition! Birds clean their feathers by preening – using their beak to comb through the strands of the feathers to make sure they are sitting neatly. Sometimes, birds spread oil from the uropygial **gland**, found near their tail, all over their feathers to help keep them waterproof and tidy.

Bald eagle

The primary wing feathers are the longest wing feathers. They are found on the outer part of the wing and push the bird forwards in the air.

The secondary wing feathers form the inner part of the wing – they push the bird up during flight.

COLOURFUL CAROTENOIDS

Birds can have feathers in a rainbow of different colours. Some of these colours are from **pigments** within the feathers, while others are just illusions, created by the structure of the feathers. Certain pigments can't be made by birds themselves – yellowy carotenoids can only be obtained by eating plants.

INTEGUMENT

FUR

In many cases, hair covers the entire body of a **mammal**, forming a coat of fur. Fur comes in a limited range of colours, and it is often grey or brown, but it can come in all kinds of patterns to help animals blend in with their surroundings.

DORMOUSE

Dormice are covered in a coat of fur — even their tail is furry. They have darker fur on their backs than on their bellies. This special pattern of colouring is called countershading and it helps to prevent them from being seen by **predators**.

Countershading

Many species of animal are countershaded. Usually, when the sun shines on an animal from above, a shadow is cast that shows that it is three-dimensional. However, a countershaded animal looks flat because its pattern balances out any shadows. This makes it harder for predators to spot it.

If the dormouse was one colour, the sun would create a shadow on its underside

A countershaded dormouse is lighter underneath

If the dormouse is countershaded its pattern counteracts any shadows

LIFE SIZE

SEA OTTER

The sea otter has the thickest fur of any animal – it is about 1,000 times denser than human hair! As well as keeping it warm, its fur also provides a waterproof layer to keep it dry.

SLOTH

Amazing sloth fur is not just a coat, it's a whole ecosystem! Algae, fungi, moths, and other insects all live within the long, coarse hairs. The algae can even make a sloth's fur look green.

PORCUPINE

The sharp and spiky shield of needle-like quills that is found on a porcupine's back is formed of extremely thick hairs. Usually the quills lie flat, but when the porcupine is alarmed it can raise them in defence.

STOAT

In chilly areas where snow falls heavily over the winter months, animals must swap their brown coats for white ones to remain **camouflaged**! When the stoat changes from brown to white, it is known as an ermine.

LLAMA

Llamas live in high mountains where it can be very cold. They have dense, two-layered fur. Fluffy hair underneath helps to keep them warm, and longer coarse hairs above help to keep them dry.

INTEGUMENT

INSIDE A SHARK

Sharks are underwater **predators** with a streamlined body, which allows them to swim speedily through the water in pursuit of their **prey**. Their incredible skin acts like chain mail, protecting them from harm, as well as helping them to move quickly. To assist them in tracking down their food, sharks also have a sneaky extra skill – the ability to detect electricity!

Sensory information is processed in the brain. Sharks have a highly sensitive sense of smell and a large portion of their brain is dedicated to processing odours.

A lateral line runs down the length of the shark's body on either side, which allows it to feel vibrations in the water.

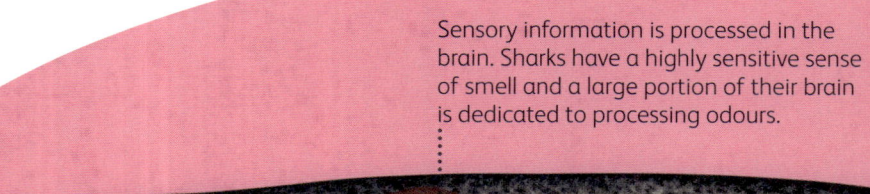

The ampullae of Lorenzini are extremely sensitive sensors that can detect the faintest of electrical signals. They are found all around the shark's snout.

Nerves take signals from the ampullae of Lorenzini to the brain so the shark can sense the electrical signals of prey.

AMPULLAE OF LORENZINI

The **hearts** and **muscles** of all animals give off weak electrical signals. Sharks can sense these electrical signals using sensors called the ampullae of Lorenzini, which are located around the shark's snout. Small **pores** on the skin lead to long, jelly-filled canals that transmit the electrical signals to receptors at their base, which send signals to the **brain** via nerves.

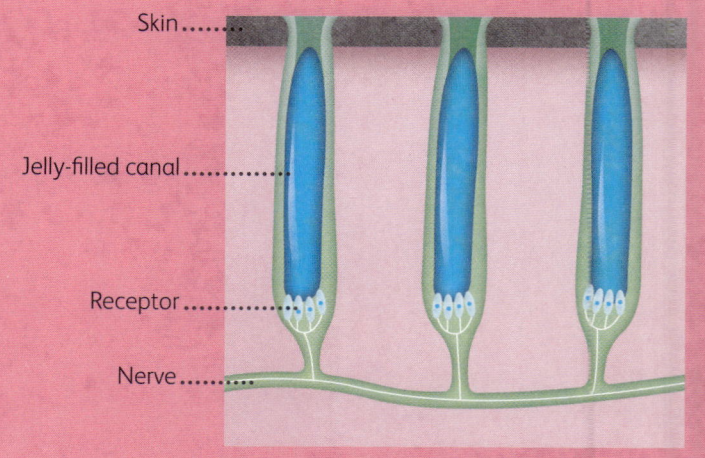

Ampullae of Lorenzini

> Unlike the scales of other fish, which grow continuously, the dermal denticles of sharks are shed and replaced with new ones.

SANDPAPER SKIN

The skin of sharks is covered in tiny, rough V-shaped scales called dermal denticles, which look and feel very different from other fish scales. Their sharp tips all point in the same direction — towards the tail — making their skin feel smooth in one direction but like sandpaper in the other.

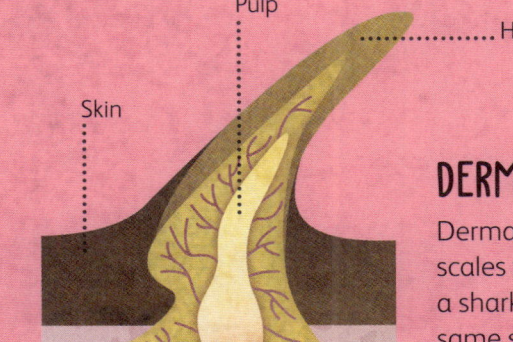

DERMAL DENTICLES

Dermal denticles are the special scales that cover the surface of a shark's skin. They have the same structure as teeth and are covered in a tough material that is a lot like **enamel**. Inside this hard exterior is a pulp cavity with a rich supply of **blood** and nerves.

Small-spotted catshark

CLEVER CAMOUFLAGE

The pattern and colouring of the small-spotted catshark's skin **camouflages** it on rocky seafloors. Like many other sharks, it also demonstrates countershading, where the top of its body is dark and the bottom is light, to help it blend in from above and below when the sun shines on it.

INVERTEBRATES

WHAT IS AN INVERTEBRATE?

Animals that don't have a **backbone** inside their body are called **invertebrates**. Some – such as slugs, worms, and octopuses – have soft bodies, with no hard **skeletons** at all. Others have a strong covering on the outside of their body known as an exoskeleton. Invertebrates are found all over the world, from the blistering dunes of the hottest deserts to the darkest trenches of the ocean.

TYPES OF INVERTEBRATE

The vast majority of animals on Earth are invertebrates. Just as **vertebrates** are classified into groups to help us better understand them, invertebrates are classified into groups too.

Echinoderms
Spiny-cased ocean creatures like starfish and sea urchins belong to the echinoderm group. They are often found on the seafloor.

Cnidarians
Cnidarians are radially symmetrical animals that live in the water. They have stinging tentacles for catching their **prey**.

Sponges
Sponges are simple aquatic creatures that have been around for 600 million years. They cannot move around.

Roundworms
These round-bodied worms are often **parasites** that live inside the bodies of other creatures. There are thousands of species.

Flat worms
These flat, soft-bodied worms often live in the bodies of other animals as parasites. Some are brightly coloured.

Segmented worms
Worms with soft bodies made of visible segments, such as earthworms and leeches, are called segmented worms.

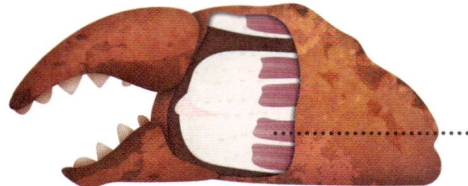

Muscles attach to hard sheets on the inside of the exoskeleton called apodemes.

EXOSKELETONS

Arthropods have external skeletons called exoskeletons, which are made of a hard material called chitin. **Muscles** attached to folds on the inner surface of an exoskeleton, called apodemes, allow arthropods to make complex movements.

FILLED WITH WATER

Some soft-bodied invertebrates, such as worms, use muscles, their body wall, and the pressure of fluid within their body to help them move instead of an exoskeleton. This is known as a hydrostatic skeleton.

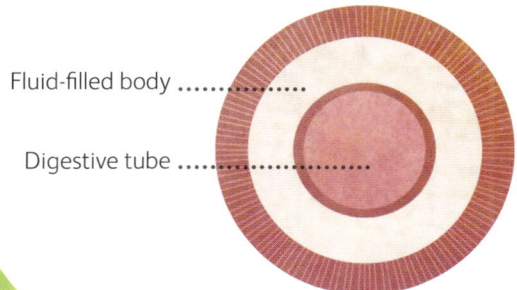

Fluid-filled body

Digestive tube

Cross-section of a worm

Molluscs

Molluscs are a huge group of soft-bodied invertebrates that includes octopuses, snails, and bivalves. Many have hard outer shells for protection.

Octopuses are highly intelligent molluscs that are excellent escape artists and can even solve complex tasks.

Snails are slime-producing molluscs that usually have spiral shells. They are found both in water and on land.

Bivalves have a hinged shell made of two parts. They are often eaten as shellfish, such as mussels and oysters.

Arthropods

The arthropod group contains most of the invertebrates we are familiar with, such as insects and spiders. They have hard exoskeletons, jointed legs, and bodies that are split into segments.

Six-legged insects have bodies split into three sections. This group includes ants, butterflies, cockroaches, bees, and grasshoppers.

Crustaceans are mostly found in the sea, such as crabs, lobsters, and shrimp. Many have a pair of snapping claws.

Centipedes and millipedes are arthropods with many legs, and long, thin bodies made up of lots of segments.

Arachnids have bodies split into two parts and eight legs. This group includes spiders, ticks, scorpions, and mites.

INVERTEBRATES

COMPOUND EYES

Up close, damselflies look like aliens. This is partly due to their huge compound eyes, which wrap around their head. While **vertebrates** usually have just two eyes, damselflies have thousands. Each of the damselfly's compound eyes is actually made up of many smaller eyes, known as ommatidia. One damselfly can have around 30,000 ommatidia to help it see the world!

Antennae are used for sensing touch and smell.

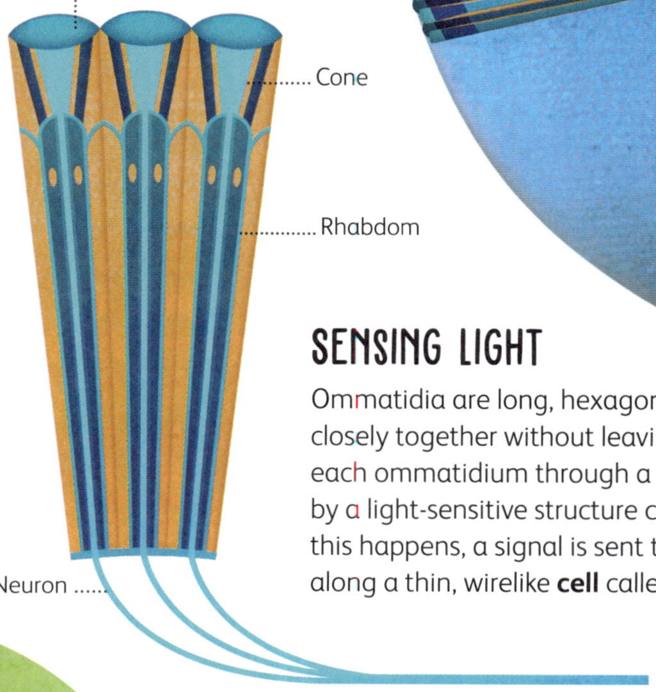

Clear cap

Cone

Rhabdom

Neuron

SENSING LIGHT

Ommatidia are long, hexagonal tubes, which pack closely together without leaving a gap. Light enters each ommatidium through a clear cap and is detected by a light-sensitive structure called a rhabdom. When this happens, a signal is sent to the insect's **brain** along a thin, wirelike **cell** called a neuron.

ULTIMATE HUNTER

Damselflies and their close relatives dragonflies are superb hunters. They zoom through the air, using basket-like legs to snatch up their insect prey. Their large eyes give them excellent all-round vision and they can use their four wings to manoeuvre in all directions.

Damselflies have three extra eyes on the top of their head called ocelli. These sense whether it is light or dark.

FALSE PUPIL

It might look like this praying mantis is staring right at you, but the black dots in its eyes aren't pupils. You are actually staring down the ommatidia that are directly facing you, which absorb all the light that enters them, making them appear dark.

Common blue damselfly

Unlike other animals, insects cannot rotate their eyeballs. To see around them, they must move their whole body or head.

INSECT VISION

Each ommatidium acts like an individual eye, capturing small sections of an image. When all of these images are put together in the damselfly's brain, it may give a pixelated view of the world like the one pictured below.

INVERTEBRATES

SPIRACLES

In **vertebrates**, **oxygen** is carried to the **organs** in **blood**. Insects, however, use an almost totally different system! A network of tiny tubes guide air from outside the body directly to where it is needed. Because of the thick exoskeletons of many insects, air cannot move directly through their skin. Instead, it enters the body through a series of small holes, which can be opened and closed to prevent water loss.

All insects, millipedes, and centipedes have spiracles.

SPIRACLES

The holes that allow air to enter an insect's body are called spiracles. They can be seen on the outside of the abdomen as a row of black dots. Spiracles are controlled by **muscles** and connect to tubes called tracheae.

Handsome cross grasshopper

Spiracles are like small **valves** that open and close to let air in and out of the body.

The largest tracheae run inwards from the spiracles and along the length of the body.

Tracheole   Tracheae

Spiracle

BOOK LUNGS

Some spiders and their relatives have spiracles, but others have a different system for getting oxygen. They use organs inside their abdomen called book lungs, which are made up of lots of thin sheets of **tissue**, a bit like the pages of a book. Gases pass between the air and the circulatory system across these sheets.

BRANCHING NETWORK

A branching network of tubes called tracheae carry gases to every part of the insect's body. Spiral rings of a strong material called chitin prevent them from collapsing. At the end of each branch a tiny, very thin-walled tube called a tracheole connects the airways to individual **cells**.

MOULTING

As insects grow bigger they must replace their old exoskeleton with a brand new one. During this process, known as moulting, they use **air sacs** inside their body to expand their new skin and break free of the old one. During the moult, the spiracles and lining of the tracheae are also fully replaced!

In larger insects, thin-walled sacs fill with air to help pump gas through the tubes.

PUPA

Some insects start their life looking one way, before turning into an adult that looks totally different! When insects undergo a radical change like this we call it complete metamorphosis. In this kind of life cycle the stage that hatches out of the egg is known as a larva. The change from larva to adult happens inside a special casing called a pupa. In butterflies, a larva is known as a caterpillar and a pupa is known as a chrysalis.

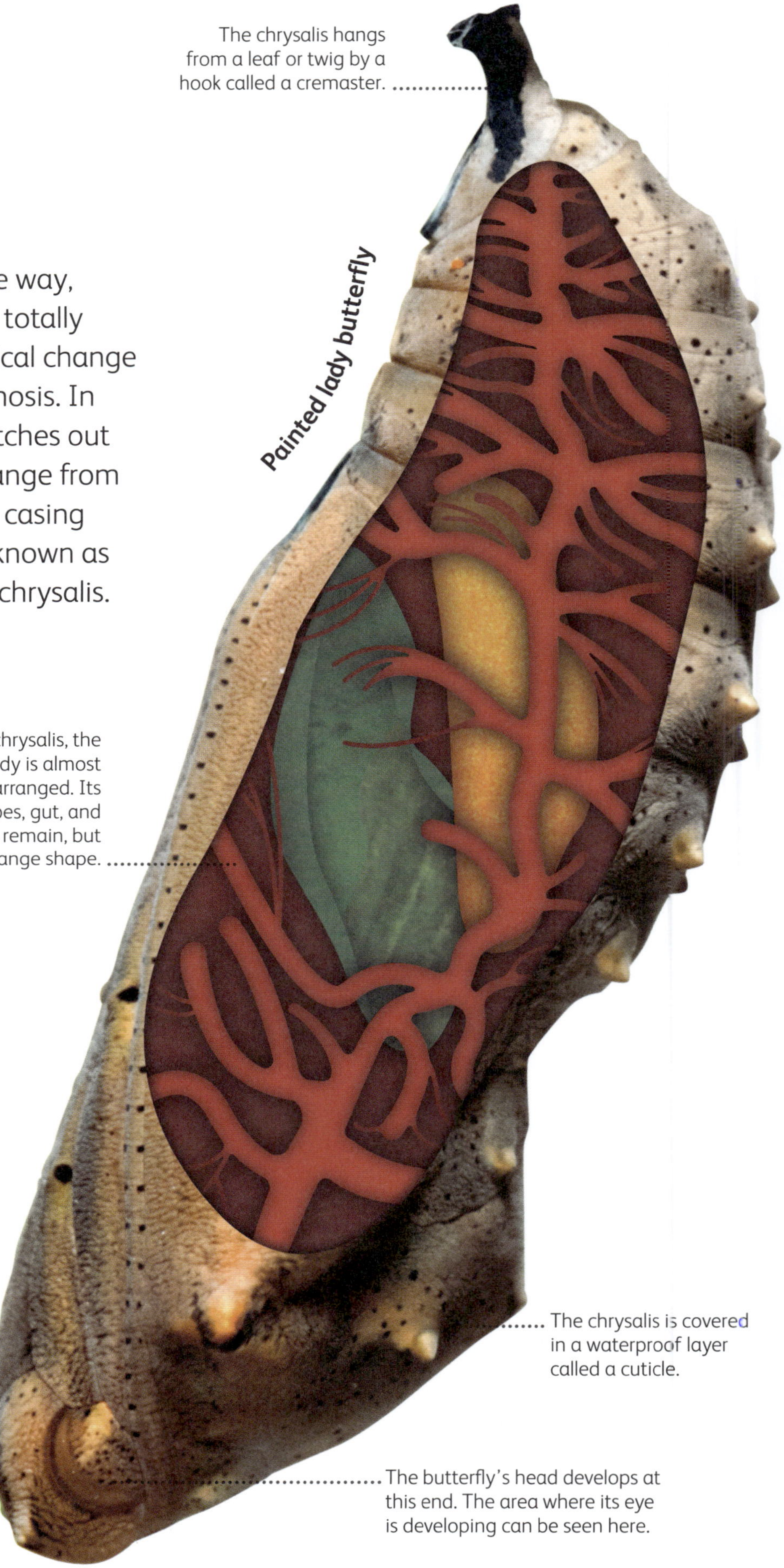

The chrysalis hangs from a leaf or twig by a hook called a cremaster.

Painted lady butterfly

Inside the chrysalis, the caterpillar's body is almost completely rearranged. Its breathing tubes, gut, and **air sacs** remain, but change shape.

The chrysalis is covered in a waterproof layer called a cuticle.

The butterfly's head develops at this end. The area where its eye is developing can be seen here.

CHANGING CATERPILLAR

When a caterpillar hatches from its egg, it is very small. As it grows bigger, it moults its old exoskeleton, growing a new one underneath. Most caterpillars moult four or five times before turning into a pupa — we call each of these stages an instar. Each instar looks slightly different to the one before.

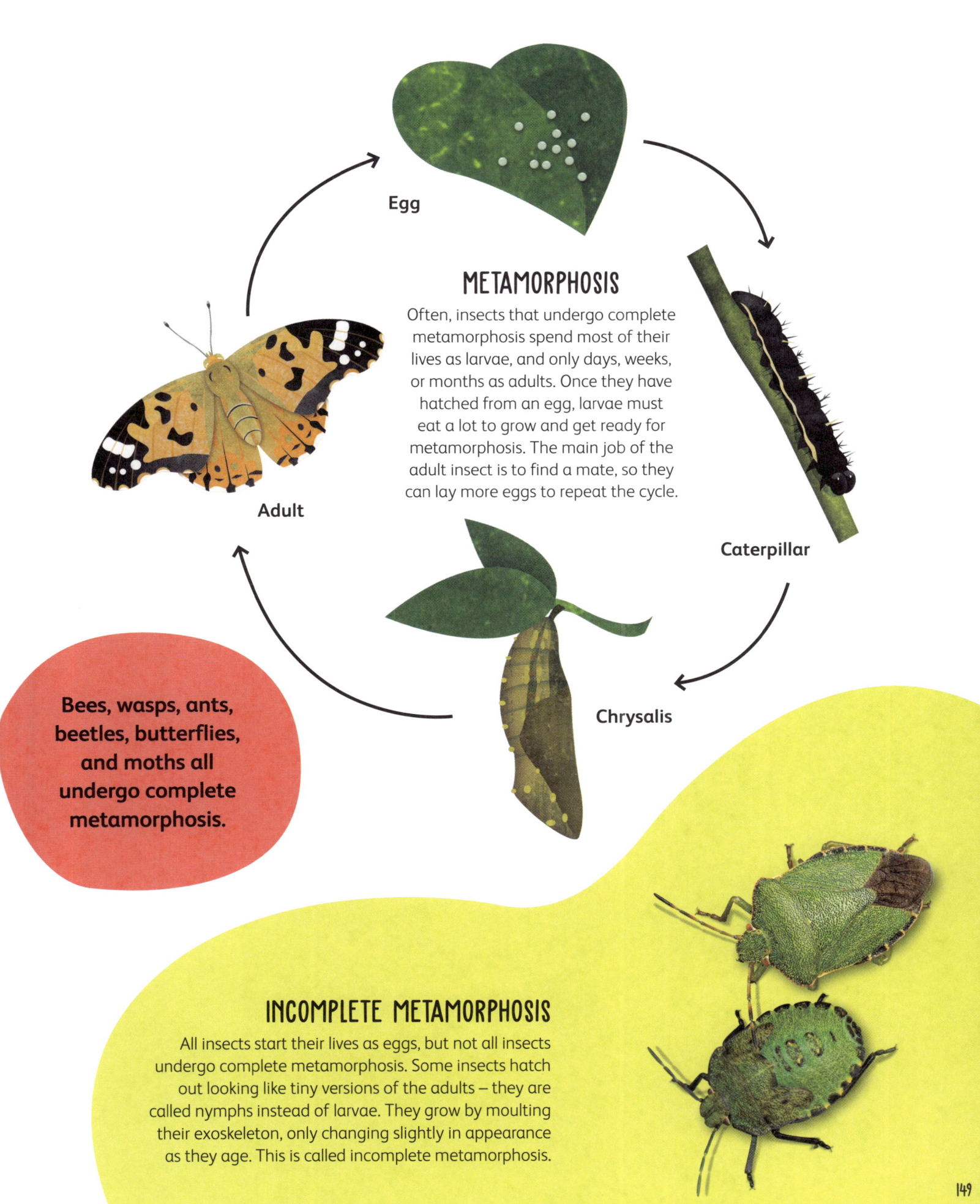

Egg

Caterpillar

Chrysalis

Adult

METAMORPHOSIS

Often, insects that undergo complete metamorphosis spend most of their lives as larvae, and only days, weeks, or months as adults. Once they have hatched from an egg, larvae must eat a lot to grow and get ready for metamorphosis. The main job of the adult insect is to find a mate, so they can lay more eggs to repeat the cycle.

Bees, wasps, ants, beetles, butterflies, and moths all undergo complete metamorphosis.

INCOMPLETE METAMORPHOSIS

All insects start their lives as eggs, but not all insects undergo complete metamorphosis. Some insects hatch out looking like tiny versions of the adults – they are called nymphs instead of larvae. They grow by moulting their exoskeleton, only changing slightly in appearance as they age. This is called incomplete metamorphosis.

INVERTEBRATES

TENTACLES

Jellyfish, anemones, and corals belong to the group of animals known as cnidarians. All of the members of this group have tentacles – long appendages that they use to catch their dinner. The tentacles of all cnidarians have stinging **cells** that contain **venom** to subdue their **prey**. Cnidarians have simple but beautiful bodies and either live attached to the seafloor or swim freely in the ocean.

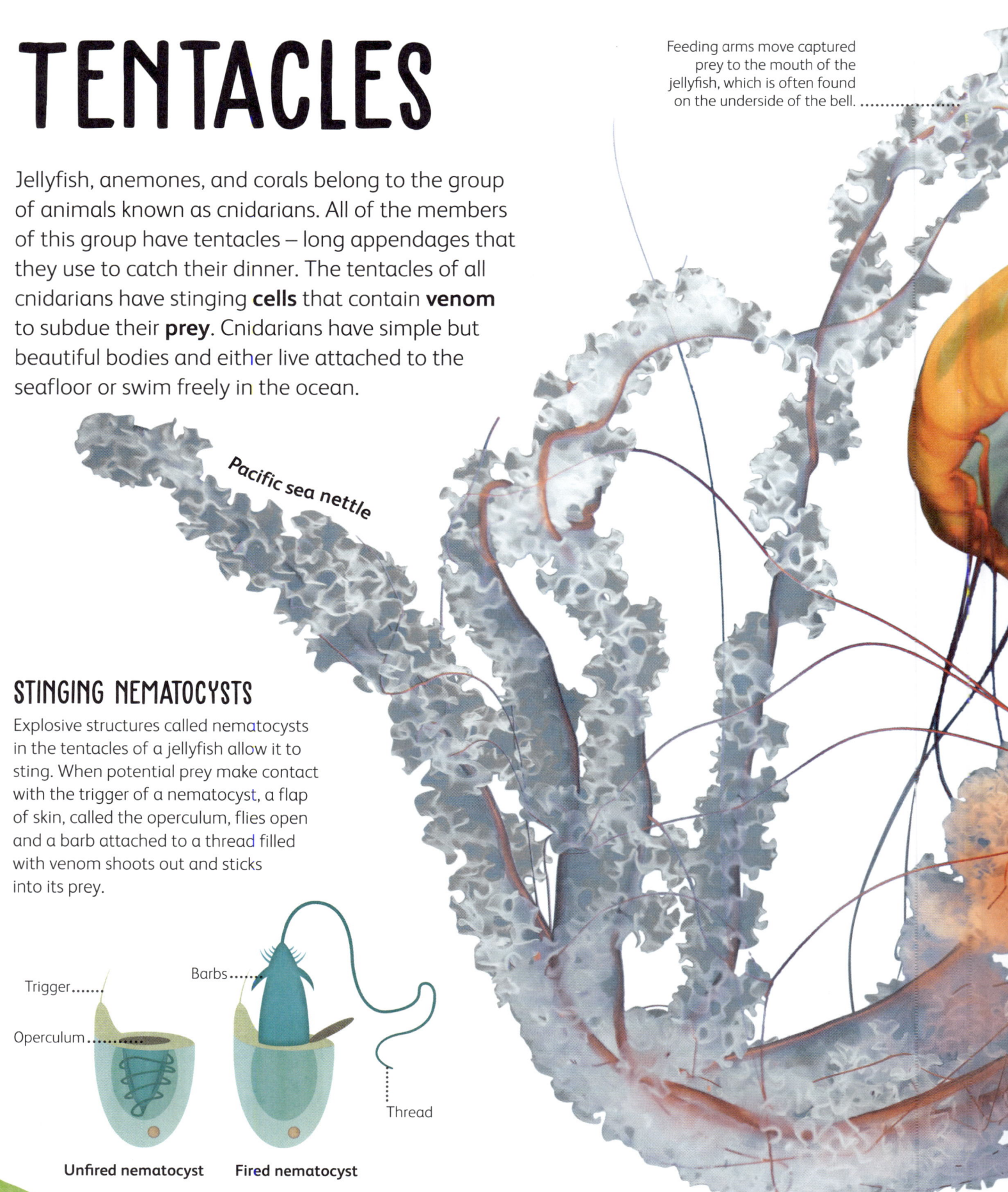

Feeding arms move captured prey to the mouth of the jellyfish, which is often found on the underside of the bell.

Pacific sea nettle

STINGING NEMATOCYSTS

Explosive structures called nematocysts in the tentacles of a jellyfish allow it to sting. When potential prey make contact with the trigger of a nematocyst, a flap of skin, called the operculum, flies open and a barb attached to a thread filled with venom shoots out and sticks into its prey.

Trigger
Operculum
Barbs
Thread

Unfired nematocyst   **Fired nematocyst**

The umbrella-like body of a jellyfish is called the bell.

The tentacles are covered in many small stinging cells.

LONG TENTACLES

The giant lion's mane jellyfish has large numbers of densely packed tentacles, which can grow to extraordinary lengths. Frequently measuring up to 3 m (10 ft) long, some individuals have been known to have tentacles measuring more than 30 m (98 ft)!

LIFE CYCLE

Jellyfish have a complex life cycle. They start their lives as eggs, which hatch into larvae that attach themselves to rocks and become polyps. A polyp clones itself to release a baby jellyfish called an ephyra, which eventually becomes a medusa, or adult jellyfish! The polyp, ephyra, and adult medusa have tentacles with nematocysts.

ARMS OR TENTACLES?

Some molluscs have tentacles too, such as the appendages that a snail uses to feel its way along. Squid have both arms and tentacles. The difference is that their eight arms have suckers all the way along them, while their two tentacles only have suckers at the end.

Moulting

As the soft insides of arthropods grow, they become too big to fit inside the exoskeleton. When this happens, the arthropod climbs out of its old exoskeleton to reveal a new, bigger one underneath! This process is called moulting.

FIDDLER CRAB

Male fiddler crabs famously have one large claw, which they use to compete with other males. If they feel threatened, fiddler crabs can detach this large claw to distract their attacker and give them time to escape. Like most arthropods, they can regrow these limbs when they moult.

LIFE SIZE

ARMOUR

Many **invertebrates** have hard armour for protection, to prevent water loss, and to provide support for their bodies. Arthropods have exoskeletons made of tough chitin. Because their exoskeleton is solid, arthropods have jointed limbs to allow them to move. Some molluscs also have hard armour in the form of shells, but these are made of a substance called calcium carbonate.

SNAIL

Snails have soft, squishy bodies that are protected by hard, spiral shells. Unlike arthropods, the shells of snails and other molluscs get bigger as they grow older.

SCALLOP

Bivalves, such as scallops, are molluscs and have shells made of calcium carbonate. The king scallop's shell is formed of two curved halves held together by a **ligament** at a hinge.

WOODLOUSE

Woodlice have exoskeletons made of overlapping plates, which allows some species to curl up into a ball. Unlike many other arthropods, woodlice moult their exoskeletons in two halves.

HERCULES BEETLE

Some arthropods use parts of their exoskeleton as weapons, such as the horns of the male Hercules beetle. These are incredibly strong and allow the beetles to lift rival males clean into the air.

SCORPION

Scorpions have a long tail and their hard exoskeleton forms a sharp, pointed stinger at the tip for delivering **venom**. Incredibly, scorpion exoskeletons glow under ultraviolet light!

INVERTEBRATES

INSIDE AN OCTOPUS

Soft-bodied octopuses are molluscs – a group of **invertebrates** whose bodies are wildly different from those of any others. Octopuses are intelligent and curious animals, which feel their way through their watery world using eight long, sucker-covered arms. When they feel threatened, octopuses can rapidly change colour, brightening up the ocean like living kaleidoscopes

CRUNCHING BEAK

Many of an octopus's favourite foods, such as crabs and clams, have hard outer coverings. To break them up into bite-size pieces, octopuses have hard, parrot-like beaks. The beak is the only hard part of an octopus's body.

The octopus's main heart pumps **oxygenated** blood around the body.

Each one of the two smaller hearts, found on either side of the octopus, pumps blood to a set of gills.

Octopus gills look like feathery combs. They are also known as ctenidia.

THREE HEARTS

Octopus **blood** contains copper instead of **iron**, meaning that when it is full of **oxygen** it is blue instead of red. The blood is pumped around the body by the main **heart**, but octopuses also have two extra hearts, which are responsible for pumping **deoxygenated**, colourless blood through the **gills**.

CHROMATOPHORES

Chromatophores are **cells** filled with **pigment** under an octopus's skin. When a ring of **muscles** pulls on each cell, the sac of pigment is exposed, making that patch of skin appear the colour of the pigment inside. The colour change happens in less than a second, so an octopus can **camouflage** itself instantly.

SUCKERS

Octopus arms are made of muscle and covered in suckers, which they use to grab objects and explore their **environment**. The suckers are covered in sensors, which can both touch and taste whatever they come into contact with! The covering of the suckers is shed as the octopus grows.

Octopus skin contains thousands of colour-changing cells called chromatophores.

Common octopus

Octopuses have eight arms. Each arm has its own mini **brain**, allowing it to taste, touch, and move without instructions.

GLOSSARY

ACID
type of strong chemical that can break down other substances

ADAPTATION
feature of an organism that helps it to survive in its environment

AIR SAC
air-filled organ found in different animals. The air sacs of birds and insects help them to breathe

AMPHIBIAN
type of animal with thin, slimy skin that lives on both land and in water

ARTERY
thick-walled blood vessel that carries blood from the heart to the body

BACKBONE
spine

BACTERIA
microscopic organisms with one cell each

BIRD
type of animal with feathers and a beak

BLOOD
liquid found in the circulatory system of animals that carries oxygen, carbon dioxide, and other substances around the body

BLOOD VESSEL
tubular organ that blood flows through. There are three types of blood vessel: arteries, veins, and capillaries

BONE
one of many hard organs that make up the skeleton of vertebrates

BONE MARROW
tissue found inside some bones that helps the bone to grow and that produces red blood cells

BRAIN
organ that receives and sends messages to the rest of the body via the nervous system

CAMOUFLAGE
colour and pattern of an organism that helps it to hide in its environment

CAPILLARY
small blood vessel that carries blood to the body

CARBON DIOXIDE
gas that is a waste product of respiration

CARNIVORE
animal that eats only meat

CARTILAGE
firm but flexible tissue that makes up parts of the body, such as the nose, and the skeletons of cartilaginous fish

CARTILAGINOUS
made of cartilage. Cartilaginous fish, including sharks and rays, have skeletons made of cartilage

CELL
smallest unit of a tissue

CHEMICAL
substance made from one or more elements

CHEMICAL REACTION
when two or more chemicals combine or split to form a new substance

CLOACA
opening found on the rear of amphibians, reptiles, and birds where the digestive, urinary, and reproductive systems exit the body

COLD-BLOODED
description of an animal that does not make its own heat and must get warmth from its environment

COLLAGEN
type of protein that is found in many tissues of the body, including the skin

CONNECTIVE TISSUE
type of tissue that sticks together or separates other tissues

DEOXYGENATED
description of blood or water with a low level of oxygen

DIAPHRAGM
large, dome-shaped muscle found in mammals below the lungs and above the liver, which helps with breathing

DIGESTION
when food is broken down in the body and the useful nutrients are extracted

ECHOLOCATION
when animals, such as certain bats, navigate in the dark by making loud calls and listening to the echoes to build up a picture of their surroundings

ENAMEL
hard tissue that makes up the outer covering of the teeth of mammals

ENVIRONMENT
habitat in which an organism lives

ENZYME
chemical that helps to break down other substances

EVOLUTION
when a species becomes better adapted to its environment over successive generations

EXCRETION
when waste products exit the body, often as urine or faeces

FAECES
poo

FAT
tissue in the body that stores energy. It is often found under the skin to help with insulation

FERMENTATION
when sugars are broken down, often by microbes

FIBRE
part of food that cannot be digested or a long cell within a muscle

FISH
type of animal with scales and gills that lives underwater

GAS EXCHANGE
when oxygen is absorbed and carbon dioxide is released by cells, so respiration can continue

GILLS
respiratory organs found in animals that live underwater. They absorb oxygen from water

GLAND
organ that produces a substance to be used in the body, such as saliva

HEART
organ that pumps blood around the body

HERBIVORE
animal that eats only plants

HORMONE
chemical made by glands in the body that tells other parts of the body to perform certain functions

IMMUNE SYSTEM
network of organs in the body that fight infections

INFRARED
type of light invisible to humans that is given off by warm objects or organisms

INTESTINES
digestive organs that extract nutrients and water from food. There are two main parts of the intestines: the small intestine and the large intestine

INVERTEBRATES
large group of animals that do not have a backbone

IRON
metal that is found in small amounts in the body, including in the blood of some animals

JAWS
bones that support the mouth. The jaws are made up of the upper jaw and the lower jaw

JOINT
where two bones meet and connect

KERATIN
type of protein that makes up hair, fingernails, horns, and hooves

LIGAMENT
cord that connects bones or organs together

LUNGS
respiratory organs usually found in animals that live on land. They absorb oxygen from air

MAMMAL
type of animal with hair that feeds its young with milk

MATE
individual of the opposite sex that an animal produces offspring with, or the process of a male transferring sperm to a female in order to fertilize her eggs

MELANIN
pigment that makes skin or hair darker

MEMBRANE
thin sheet of tissue

MICROBE
microscopic organism made of a single cell

MIGRATION
when an animal moves over a large distance, usually during a particular season, to breed or find food

MINERAL
solid substance that is used by the body, such as calcium for bone strength

MUCUS
slimy gel made by animals to keep particular organs moist

MUSCLE
organ that contracts or relaxes to make an animal move

NUTRIENTS
useful substances in food that are extracted during digestion to help the body grow, repair, and maintain itself

OESOPHAGUS
tubular organ that connects the back of the mouth to the stomach

OMNIVORE
animal that eats both meat and plants

ORGAN
part of the body that performs a certain function, such as the heart, which pumps blood around the body

OXYGEN
gas needed for respiration

OXYGENATED
description of blood or water with a high level of oxygen

PAIRED MUSCLES
two muscles that act in opposing directions to move one part of the body

PARASITE
organism that steals food from another to survive

PATHOGEN
microbe that causes disease

PIGMENT
colourful chemical

PORE
small opening in the skin

PREDATOR
animal that hunts other animals

PREHENSILE
able to grasp

PREY
animal that is hunted by other animals

PROTEIN
type of substance that makes up many organs in the body, including muscles

REPTILE
type of animal with scales and dry skin

RESPIRATION
when cells create energy from a chemical reaction using oxygen and glucose

RUMINANT
type of herbivore with a multi-chambered stomach that digests its food twice

SALIVA
watery mucus produced in the mouth of animals to help them digest food and kill pathogens

SKELETON
hard internal frame of vertebrate animals made of bones

SKULL
bones of the head that protect the brain

SPECIES
particular type of plant or animal, such as a tiger

SPINAL CORD
bundle of nerves found inside the backbone that is a key part of the nervous system

SUGAR
type of substance that is full of energy

TENDON
cord that connects bones to muscles

TISSUE
group of cells that performs the same job. Tissues make up organs

TOXIN
substance that is harmful to an organism

TRACHEA
tubular organ that connects the back of the mouth and nostrils to the lungs, also called the windpipe

URINE
watery fluid produced by the kidneys to remove waste products from the blood

VALVE
opening inside blood vessels that can be closed to prevent blood from flowing backwards

VEIN
thin-walled blood vessel that carries blood from the body back to the heart

VENOM
toxic substance that causes harm when injected into an animal, for example by fangs or a stinger

VERTEBRATES
group of animals that have a backbone

VITAMINS
nutrients needed by the body to perform different functions

INDEX

A
adaptations 30, 42, 44, 54, 58, 67, 72, 76, 85, 101, 106
air sacs 26, 66, 67, 72
alveoli 63, 69
amphibians 10, 38–39, 49, 62–63, 111, 112–113, 130–131, 133
ampullae of Lorenzini 138
antlers 22–23
anus 77, 85
aposematism 130
Arctic terns 66–67
armour 152–153
arteries 48, 49, 51
arthropods 143, 153
axolotl 62
axons 95

B
backbones 18, 24, 95
baculum 119
basking 57
bats 30–31
beaks 11, 78, 90, 127, 135, 154
bears 82–83
beavers 89
beetles 153
binocular vision 97
birds 10–11, 26–27, 29, 36–37, 39, 49, 66–67, 77, 90–91, 100–101, 104, 105, 111, 114–115, 120–121, 127, 134–135
bison 80–81
bivalves 143, 153
bladder 87
blood 19, 35, 48–59, 62, 63, 64, 65, 68, 69, 86, 87
blood vessels 48, 49, 50, 52, 56, 126
blowholes 68
blubber 128–129
bones 18–31
book lungs 147
brain 22, 58, 94, 95, 98, 101, 138
brown fat 129
butterflies 148–149

C
camels 54–55
camouflage 44, 137, 139
capillaries 49, 53, 62, 65
carapace 24
carbon dioxide 63, 69
cardiac muscle 35
carotenoids 135
cartilage 20–21
cartilaginous fish 20–21
caterpillars 148, 149
cats 96–97
cervix 117
chambers, heart 51
chameleons 29, 44–45
changes, body 9, 113, 148, 149
chewing 40–41, 81
chromatophores 155
chrysalises 148, 149
circulation 12, 48–59
claws 29, 127
cloaca 77, 87, 111, 114, 123
cnidarians 142, 150–151
collagen 38
columella 101
common ancestor 15
comparative anatomy 14–15
compound eyes 144–145
convergent evolution 15
countercurrent heat exchange 56
countershading 136
courtship 120–121
coyotes 71
crabs 143, 152
crocodiles 25, 133
crop 90
crustaceans 143, 152

D
damselflies 144–145
dermal denticles 139
dermis 126, 132
diaphragm 66, 73
digestion 13, 76–91
dissection 6
dolphins 14–15, 37, 71, 105
dormice 136
down 134

E
eagles 29
ears 23, 31, 94, 100–101, 107, 133
echinoderms 142
echolocation 31, 105
eggs 110, 112–115, 117, 149, 151
electroreception 104, 138
elephants 43, 89, 106–107, 116
enzymes 77, 81, 83, 85, 91
epidermis 126, 132
evolution 14, 15
exoskeletons 143, 146, 152, 153
eyelids 97
eyes 45, 94, 96–97, 133, 144–145

F
facial disc 100
facial expressions 41
faeces (poo) 84, 85
fallopian tubes 111, 117
fat 128–129
feathers 126, 127, 134–135
feet 106, 107, 133
female anatomy 111, 116–117
fennec foxes 94–95
fertilization 111
filter feeding 78
fingers 29, 30
fish 10, 20–21, 49, 52–53, 62, 64–65, 89, 105, 111, 112, 113, 121, 127, 132, 138–139
flatworms 142
flehmen response 99
flight 26, 30–31, 36–37
flippers 29
follicles 103
fossils 9
frogs 6–7, 38–39, 70, 87, 112, 121, 130–131
fur 136–137
fused bones 24–25

G
gallbladder 83
gannets 67
gastralia 25
gastroliths 91
geckos 132–133
gestation period 116, 122
gills 13, 52, 62, 64–65, 154
giraffes 42–43, 58–59
gizzard 91
glottis 72
gorillas 40–41
grasshoppers 146–147
gular fluttering 57

H
habitats 14
hagfish 53
hair 126, 127, 136–137
hands 28–29
heart 48, 49, 50–51, 59, 154
hibernation 83, 129
hippos 126
hooves 28
hormones 111, 118
horns 127, 153
horses 28, 34
hummingbirds 36–37
hunting 44, 68, 97, 100, 101, 102, 145
hydration 54
hydrostatic skeletons 143
hypodermis 126, 129

I
imaging 7
infrared sense 104
insects 143, 144–149
integument 13, 126–139
intestines 77, 80, 84–85
invertebrates 11, 13, 142–155

J
jaws 21, 22, 23, 41
jellyfish 79, 150–151
jerboas 101
joints 19, 20, 38

K
kangaroos 122–123
keratin 25, 78, 9, 103, 127, 132, 134

kidneys 86–87
kinkajous 42
kiwis 114–115
knees 31
koalas 84

L
lampreys 89
larvae 148, 149, 151
lateral lines 105, 138
ligaments 18
lions 41, 110–111
lips 42, 99
liver 77, 82–83
lizards 132–133
llamas 137
lungs 26, 48, 63, 65, 66–69, 72

M
magnetoreception 104
male anatomy 118–119
mammals 11, 30–31, 40–41, 49, 50–51, 54–55, 58–59, 68–69, 76–77, 80–85, 94–99, 101, 102–103, 106–107, 110–111, 114, 116–119, 122–123, 127, 128–129, 136–137
mammary glands 116, 117
manatees 42
marine iguanas 86–87
mating 110, 112, 120–121
metamorphosis 113, 148, 149
migration 67, 104
milk 116, 122, 128
moles 29
molluscs 143, 151, 153, 154–155
monkeys 29, 121
moulting 147, 152
mouth 76, 78–79
mucus 44, 98, 99, 131
muscles 12, 34–45, 73, 80, 143

N
narwhals 89
nematocysts 150, 151
nephrons 87
nerves 94, 95, 96, 126, 138
nervous system 95
neurons 95
noses 22, 94, 98–99, 106
nostrils 63, 98
nymphs 149

O
octopuses 143, 154–155
oesophagus 76, 79, 80, 90
olfactory bulb 98
ommatidia 144, 145
optic nerve 96
organ systems 12–13
osteoderms 133
otters 102–103, 137
ovaries 111, 117
owls 100–101
oxygen 48, 52, 62–63, 64, 66–69, 131, 146, 154

P
paired muscles 36–37
pancreas 77, 83
panting 57
parrots 71, 90–91
paws 29
peafowl 121
penguins 27, 56
penis 110, 119
pheromones 99
pinna 101
placenta 116, 117
plastron 24, 25
plates 126, 127, 153
platypuses 104
pneumatic bones 26–27
poison 130, 131
polyps 151
porcupines 137
pouches 122, 123
praying mantises 145
preening 135
prehensile body parts 42–43
proboscis 99, 121
proventriculus 91
pupa 148–149
pupils 96, 97, 145

R
rabbits 84–85
raccoons 76–77, 119
rats 118–119
red blood cells 54, 55
regurgitation 91
reproduction 13, 110–123
reptiles 10–11, 24–25, 44–45, 49, 72–73, 78–79, 86–87, 112, 115, 127, 132–133
respiration 12, 62–75
retina 96
ribs 18, 25, 73
rodents 103
roundworms 142
ruminants 80

S
sagittal crests 41
salt glands 86
scales 126, 127, 132–133, 139
scallops 153
scorpions 153
scutes 25, 127
sea lions 68–69
seahorses 42, 81
segmented worms 142
senses 9, 13, 94–107
sharks 14–15, 20–21, 82, 89, 113, 138–139
sheep 116–117
shells 114–115
shrews 50
skeletal muscles 35
skeletons 7, 12, 18–31
skin 13, 62, 94, 126, 130–131, 139, 155
skinks 43, 133
skulls 19, 22–23, 27, 41, 76
sloths 137
smooth muscle 35
snails 143, 153
snakes 19, 70, 72–73, 88, 99, 104, 132

sounds 70–71
sound waves 94, 100
sperm 110, 119
spiders 143, 147
spinal cord 95
spiracles 146–147
sponges 142
squid 151
stoats 137
stomach 77, 80–81, 91
suckers 154, 155
sugars 62, 63
sweat 57, 126
swim bladders 53
symmetry 8

T
tails 18, 42–43, 45
tapetum lucidum 96
tapirs 98–99
teeth 23, 76, 85, 88–89
temperature 56–57
tendons 18, 36, 38–39
tentacles 150–151
testes 110, 119
tigers 18–19, 71
toads 112–113
tongues 42–43, 44, 94
tortoises 24–25, 132
trachea 63, 66, 68, 70, 72, 79
tracheae 146, 147
tree of life 15
trunks 43, 99, 106
turtles 24, 78–79

U
ultraviolet sense 105
umbilical cord 117
urethra 110, 119
urine 86, 87
uterus 111, 116, 117, 123

V
vagina 111, 117, 119, 123
vas deferens 119
vasodilation 57
veins 48, 49, 58
venom 88, 150
vertebrates 10–11
vets 7
vocal cords 70
vomeronasal organ 98, 99

W
walruses 128–129
whales 29, 50–51, 68, 78, 105
whiskers 94, 102–103
white blood cells 55
wings 29, 30–31
woodlice 153
woodpeckers 27
worms 142, 143

ACKNOWLEDGEMENTS

DK would like to thank: Abi Maxwell for editorial assistance; Eleanor Bates, Ray Bryant, and Sif Nørskov for design assistance; and Clare Lloyd and Sophie Parkes for proofreading.

The publisher would like to thank the following for their kind permission to reproduce their photographs: (Key: a-above; b-below/bottom; c-centre; f-far; l-left; r-right; t-top)

1 Alamy Stock Photo: Panoramic Images. **6 Dreamstime.com:** Dragan Andrii (br). **7 Dreamstime.com:** Isselee (tc); Ivonne Wierink (tr). **Science Photo Library:** (crb); ARIE VAN 'T RIET (bc). **8 Alamy Stock Photo:** Nature Photographers Ltd / PAUL R. STERRY (tl). **Dreamstime.com:** Isselee (r). **9 Alamy Stock Photo:** ReneTi (cl); Nobumichi Tamura / Stocktrek Images (tr). **Dreamstime.com:** Kyslynskyy (c); Irina Pislari (br). **10 Dreamstime.com:** Kamensky (crb). **Science Photo Library:** Georgette Douwma (cl). **10-11 Dreamstime.com:** Isselee (ca). **11 Dreamstime.com:** Adrian Eugen Ciobaniuc (clb); Isselee (tr); Puripat Penpun (crb). **13 Dreamstime.com:** Carol Buchanan. **14 Dreamstime.com:** Tony Campbell (clb). **14-15 Alamy Stock Photo:** SeaTops (b); www.allievi-photography.com (c). **16-17 Dreamstime.com:** Andrei Samkov. **18-19 SkullsUnlimited.com:** (c). **19 Dorling Kindersley:** Colin Keates / Natural History Museum, London (crb). **20 Alamy Stock Photo:** Gary Corbett (bc). **20-21 Alamy Stock Photo:** Miguel Lasa / Steve Bloom Images (t). **21 Alamy Stock Photo:** WaterFrame_mus (tr). **Science Photo Library:** Michelle Gilbert, PHD (bl). **22 Alamy Stock Photo:** Arterra Picture Library / Arndt Sven-Erik (cl); Cindy Hopkins (crb). **Shutterstock.com:** Michaelparkart (tl). **22-23 SuperStock:** Cathy Hart / Alaska Stock - Design Pics. **24-25 Dreamstime.com:** Aekkaphum Warawiang (c). **25 Dreamstime.com:** Cem Ekiztas (br). **Shutterstock.com:** El Arquitecto de Huesos (tr). **26 Alamy Stock Photo:** Steve Gschmeissner / Science Photo Library (clb). **26-27 Dreamstime.com:** Ondej Prosick (c). **27 Dreamstime.com:** Eng101 (tr); Martinmark (crb). **28 Dreamstime.com:** Alexia Khruscheva. **29 Alamy Stock Photo:** blickwinkel / fotototo (c); Cultura Creative RF / Steve Woods Photography (cla); Henri Koskinen (tc); blickwinkel / AGAMI / O. Diez (bl). **Trans-Americas Journey:** Eric Mohl (cb). **30-31 Alamy Stock Photo:** blickwinkel / AGAMI / T. Douma (c). **31 Science Photo Library:** B. G Thomson (br). **32-33 Dreamstime.com:** Isselee. **36 Dreamstime.com:** Ondej Prosick (cla). **36-37 Alamy Stock Photo:** Panoramic Images (c). **37 Dreamstime.com:** Caan2gobelow (br). **Getty Images / iStock:** OGphoto (c). **38 Shutterstock.com:** Dr Morley Read (tl). **38-39 Alamy Stock Photo:** David Cook / blueshiftstudios (c). **39 Alamy Stock Photo:** Nature Picture Library / Dale Sutton / 2020VISION (tr). **Dreamstime.com:** Meepoohya (br). **40-41 Alamy Stock Photo:** Arterra Picture Library / Clement Philippe. **41 Alamy Stock Photo:** Peter van Evert (tl). **Dreamstime.com:** Digistockpix (crb); Wirestock (tr). **42 Dreamstime.com:** Voislav Kolevski (clb); Matthijs Kuijpers (bl). **Getty Images:** Phil Lowe / 500px (crb). **42-43 Depositphotos Inc:** lifeonwhite (t). **43 Getty Images / iStock:** Image Source (bl). **44 Alamy Stock Photo:** Gillian Pullinger (tr). **45 Getty Images / iStock:** E+ / SensorSpot (tc). **46-47 Alamy Stock Photo:** Morgan Trimble. **50 Alamy Stock Photo:** Francois Gohier / VWPics (cl). **naturepl.com:** Dietmar Nill (bl). **50-51 Alamy Stock Photo:** WaterFrame_fba (c). **52 Alamy Stock Photo:** Nature Picture Library / Wild Wonders of Europe / Lundgren (bl). **52-53 Alamy Stock Photo:** blickwinkel / A. Hartl (c). **53 Alamy Stock Photo:** Mark Conlin (tr). **54 Alamy Stock Photo:** Mauricio Abreu (cl). **54-55 123RF.com:** ferli (c). **55 Alamy Stock Photo:** Nature Picture Library / Doug Allan (tr). **Shutterstock.com:** Pee Paew (crb). **56 Dreamstime.com:** Frank Gnther. **57 Alamy Stock Photo:** blickwinkel / Pieper (bc); David Osborn (br). **Dreamstime.com:** Jmrocek (cla). **Getty Images:** Luke Sharrett / Bloomberg (c). **Science Photo Library:** Peter Chadwick (tr). **58-59 Dreamstime.com:** Ecophoto. **59 Courtesy NASA/JPL-Caltech** (tr). **60-61 Alamy Stock Photo:** Life on white. **64 Alamy Stock Photo:** Mark Conlin (tl). **64-65 naturepl.com:** Alex Mustard (c). **65 Dreamstime.com:** Nic9899 (tr). **naturepl.com:** David Fleetham (cra). **66-67 naturepl.com:** Niall Benvie (c). **67 Alamy Stock Photo:** Avico Ltd (cra). **Dreamstime.com:** Altaoosthuizen (tl). **68 Alamy Stock Photo:** Nature Picture Library / Alex Mustard (cl). **Dreamstime.com:** Mikhail Laptev (tc). **69 Dreamstime.com:** Daniel Lacy (tr). **70 Alamy Stock Photo:** Eric Nathan (r). **Science Photo Library:** Dante Fenolio (clb). **71 Dreamstime.com:** Nilanjan Bhattacharya (br); Slowmotiongli (cla); Izanbar (tr); Moose Henderson (bl). **72 Alamy Stock Photo:** Joseph Salvoni (cl). **72-73 Alamy Stock Photo:** Redmond Durrell (c). **73 Getty Images:** Moment / Paul Grace Photography Somersham (bc). **74-75 Alamy Stock Photo:** Paul Ives. **78 Alamy Stock Photo:** Karen Debler (b). **Dreamstime.com:** Henner Damke (tr). **79 Alamy Stock Photo:** robertharding / Michael Nolan (b). **Getty Images / iStock:** richcarey (tr). **80 Shutterstock.com:** Wirestock Creators (bl). **80-81 Getty Images:** Suntisak P Chakorn / 500px. **81 Alamy Stock Photo:** Oleg Belov (tr). **Dreamstime.com:** Wrangel (crb). **82 Dreamstime.com:** Leopoldo Palomba (bl). **83 Alamy Stock Photo:** Vince Burton / NIS / Minden Pictures (tr). **84 Dreamstime.com:** Blasch (tl); Jeep2499 (bl). **84-85 Alamy Stock Photo:** Brian Kushner. **85 Alamy Stock Photo:** Helen Sessions (tc). **86 Dreamstime.com:** Martinmark (tl). **86-87 Alamy Stock Photo:** Barry Soames. **87 Alamy Stock Photo:** blickwinkel / Trapp (tl). **88 Alamy Stock Photo:** Pete Oxford / Minden Pictures (c). **89 Alamy Stock Photo:** Buiten-Beeld / Jelger Herder (tr); Robert McGouey / Wildlife (clb); Yvette Cardozo (c); Flip Nicklin / Minden Pictures (b). **Dreamstime.com:** Prayuth Gerabun (tl). **90 Alamy Stock Photo:** imageBROKER / P. Wegner (tl). **90-91 Dreamstime.com:** Isselee (c). **91 Alamy Stock Photo:** Media Drum World (tc). **92-93 Dreamstime.com:** Isselee. **96 Alamy Stock Photo:** imageBROKER / BA-Geduldig (tr); Zoonar / Sergey Taran (l). **97 Alamy Stock Photo:** Jacques Julien (cr). **Getty Images:** Westend61 / Creativ Studio Heinemann (tr). **98-99 Alamy Stock Photo:** VPC Animals Photo. **99 Alamy Stock Photo:** Imagebroker / Arco Images / Therin-Weise (tl); Ben Queenborough (cr); Nature Photographers Ltd / PAUL R. STERRY (br). **100 Alamy Stock Photo:** imageBROKER / Franz Christoph Robiller (tr); Harold Stiver (c). **100-101 Alamy Stock Photo:** FLPA (b). **101 naturepl.com:** Klein & Hubert (tr). **102 Alamy Stock Photo:** Nature Picture Library / Eric Baccega (clb). **102-103 Shutterstock.com:** Brockley98 (b). **103 Dreamstime.com:** Rudmer Zwerver (tr). **104 Alamy Stock Photo:** Rinus Baak (cra); Slowmotiongli (tl); Snappylens (tc). **105 Alamy Stock Photo:** imageBROKER.com GmbH & Co. KG / Andrey Nekrasov (cla). **Dreamstime.com:** Vlad Ghiea (cra); Jessamine (bl). **106 Alamy Stock Photo:** AfriPics.com (clb). **106-107 Alamy Stock Photo:** Thomas Lenne (c). **107 Alamy Stock Photo:** NSP-RF (tl). **108-109 Alamy Stock Photo:** Nature Photographers Ltd / Paul R. Sterry (b). **112 Alamy Stock Photo:** imageBROKER / Franz Christoph Robiller (tr). **112-113 Depositphotos Inc:** Hintau_Aliaksey (b). **113 Alamy Stock Photo:** Eyal Bartov (tr); Bruno Manunza (bc). **114-115 Alamy Stock Photo:** imageBROKER.com GmbH & Co. KG / Joerg Reuther (t). **115 National Museum of New Zealand Te Papa Tongarewa:** OR.016582 (cr). **116 Alamy Stock Photo:** Mark Salter (bl). **Dreamstime.com:** Patrick Gosling (b). **116-117 Dreamstime.com:** Isselee (c). **118 Shutterstock.com:** vprotastchik (tl). **118-119 Alamy Stock Photo:** Tierfotoagentur / R. Richter (b). **119 Shutterstock.com:** HappySloth (tc). **120 naturepl.com:** Dr. Axel Gebauer (cl). **121 Alamy Stock Photo:** Arterra Picture Library / Keirsebilck Patrick (br); Paul Kennedy (tc); Michael Williams (cla); Dinodia Photos RM (cra); Design Pics Inc / Thomas C. Kline, Jr. (clb). **122 Alamy Stock Photo:** Joe Blossom (bl). **naturepl.com:** Martin Willis (tr). **122-123 Shutterstock.com:** Danny Ye. **123 naturepl.com:** Mitsuaki Iwago (crb). **124-125 Dreamstime.com:** Lldd11. **128 Alamy Stock Photo:** Steven J. Kazlowski (bl). **Dreamstime.com:** Jocrebbin (tl). **128-129 Alamy Stock Photo:** Steven J. Kazlowski (c). **129 Alamy Stock Photo:** Andy Harmer (crb). **130-131 Alamy Stock Photo:** Isselee (b). **130 Alamy Stock Photo:** Pete Oxford / Minden Pictures (tr); Carver Mostardi (clb). **131 Alamy Stock Photo:** Mark Moffett / Minden Pictures (tr). **132 Alamy Stock Photo:** RooM the Agency / dikkyoesin1 (tc). **133 Alamy Stock Photo:** robertharding / Michael Nolan (tc). **134 Alamy Stock Photo:** Design Pics Inc / Ken Baehr (cla). **134-135 Getty Images:** Moment / Gary Samples (c). **135 Alamy Stock Photo:** Peter Cook (crb); Jim Jackson (tr). **136 Alamy Stock Photo:** Zoonar / Jrgen Vogt (l). **137 Alamy Stock Photo:** Farlap (br); Paulette Sinclair (cla); Urs Hauenstein (tc); Chris Mattison (cr); imageBROKER.com GmbH & Co. KG / Bernd Zoller (bl). **138-139 BluePlanetArchive.com:** Espen Rekdal (cb). **139 naturepl.com:** Sue Daly (bl). **Science Photo Library:** Dr Keith Wheeler (tl). **140-141 Science Photo Library:** Adrian Bicker. **144 Alamy Stock Photo:** blickwinkel / F. Hecker (tr). **144-145 Alamy Stock Photo:** REDA &CO srl / AGP (b). **145 Dreamstime.com:** Ecophoto (tr). **Science Photo Library:** Eye Of Science (bc, br). **146 123RF.com:** creativenature (cl). **146-147 Alamy Stock Photo:** BIOSPHOTO / Michel Gunther (b). **147 Alamy Stock Photo:** Arya Satya (cr). **Dorling Kindersley:** Liberty's Owl, Raptor and Reptile Centre, Hampshire, UK (tr). **148 Alamy Stock Photo:** blickwinkel / Hecker (clb); Dominic Robinson (r). **149 Alamy Stock Photo:** John Cancalosi (br). **150-151 Alamy Stock Photo:** Leonid Serebrennikov (c). **151 Alamy Stock Photo:** BIOSPHOTO / Tobias Bernhard Raff (br). **naturepl.com:** Nick Hawkins (tr). **Science Photo Library:** Ruben Duro (cra). **152 Alamy Stock Photo:** blickwinkel / F. Teigler (c). **153 Alamy Stock Photo:** Ingo Arndt / Minden Pictures (tr); Nature Photographers Ltd / Paul R. Sterry (cra). **Fotolia:** Eric Isselee (br). **Robert Harding Picture Library:** Marevision (tl). **155 Alamy Stock Photo:** Sergio Hanquet / Biosphoto (tr)

Cover images: Front: Alamy Stock Photo: Redmond Durrell tr, Thomas Lenne cr, Panoramic Images cb, WaterFrame_fba tl; **Dreamstime.com:** Isselee cl, Brian Kushner br; **naturepl.com:** Alex Mustard bl; **Back: Alamy Stock Photo:** robertharding / Michael Nolan bl, Miguel Lasa / Steve Bloom Images ca, VPC Animals Photo tl; **Dreamstime.com:** Ecophoto br; Spine: **Dreamstime.com:** Ondej Prosick

All other images © Dorling Kindersley